REGIONAL PERSPECTIVES IN AGRICULTURAL DEVELOPMENT

About the Author

Nalini Govind has twenty years of teaching experience. Presently, she is a senior lecturer is Economics in Lakshmi Bai College, Delhi University. She completed her M.A. in Economics from the Delhi School of Economics, Delhi University, and was later awarded the degrees of M. Phil and Ph.D. by Jawaharlal Nehru University, New Delhi. She is continuing her related research in the field of Agriculture and aspects of development as part of her post-doctoral research work.

REGIONAL PERSPECTIVES IN AGRICULTURAL DEVELOPMENT

A Case Study of Wheat and Rice in Selected Regions of India

NALINI GOVIND

CONCEPT PUBLISHING COMPANY PVT. LTD.
NEW DELHI-110059

The Indian Council of Social Science Research gave the financial support towards the publication of this work. However, the responsibility for the contents of the book—facts, opinions, conclusions etc.—is entirely that of the author and not of the ICSSR.

ISBN: 978-81-7022-087-9

Edition 2025

Published and Printed by

Since 1974

Concept Publishing Company Pvt. Ltd.
A/15-16, Commercial Block, Mohan Garden,
New Delhi-110059 (India)
T: 91-11-41101460
E: publishing@conceptpub.com, *W*: www.conceptpub.com

Preface

THIS work in its revised present form grew out of my earlier study entitled "Variations in Wheat and Rice Responses to New Technology in Agriculture : A Case Study of Selected Regions in India", for which I was awarded the degree of doctor of philosophy by the Centre for Regional Development, Jawaharlal Nehru University.

The study analysis the complex association among physical, economic and Institutional factors as they influence cropping pattern and crop productivity in different regions of India. The results of the analysis based on secondary data are blended with the findings from a sample survey of land-holdings in different regions of concentration of these crops.

Systematic analysis and evaluation of agriculture as a spatial subject is relatively rare. The present study has examined the spatial aud temporal dimensions at different area levels, (National states and districts, and within them, the sample land-holdings). Such an approach provides a realistic regional framework for evaluating the potential of different crops in the country and to formulate the spatial strategies in conformity with the regional realities. The approach to regional analysis has been essentially inter-disciplinary.

The responsibility for errors and oversights is entirely mine. I am deeply indebted to Prof. L.S. Bhat of the Indian Statistical Institute for his invaluable guidance, suggestions and assistance in bringing out the work in its present form. I thank Prof. Moonis Raza of the Centre for Regional Development, J.N.U., New Delhi (now Vice-chancellor, Delhi University) and Prof. G.S. Bhalla of the Centre for Regional Development, J.N.U., New Delhi, for their encouragement, suggestions and assistance at various stages of the work. I thank Dr. Amitabh Kundu of the Centre for Regional Development, J.N.U., New Delhi for his suggestions and assistance in statistical analysis. I am indebted to Shri M.R. Rao (then senior programmer, Computer Centre, Planning Commission) for his expert assistance at all stages of computer analysis. I thank Shri A.N. Sharma of the Indian Statistical Institute for his assistance in cartography. Grateful thanks are due to Professor Ashok Mitra, appointed as consultant by the ICSSR for this study. His comments and suggestions have helped a great deal in the preparation of this work. Finally, I thank the Indian Council of Social Science Research, New Delhi, for the liberal financial assistance towards my field-travel and computer analysis expenses and also for the liberal grant towards the publication of this book.

15 August 1985 NALINI GOVIND

Contents

List of Maps

Introduction

ONE of the aims of modern agriculture is specialisation of agriculture which is dependent on the optimal cropping pattern and overall increase in per capita output. Since agriculture is a spatial subject, growth of agriculture at the national level should aim at regional specialisation as a strategy in long term planning. When we speak of modernisation of agriculture, regional specialisation is also one of the strategy in agricultural development within the constraints of physical, economic and institutional factors. Modern Agriculture is one in which,

i. Technology and efficiency of farming are continuously being improved through the use of improved machinery, pesticides, seeds and fertilisers.
ii. The commodities produced on farms are changing constantly in response to changes in consumer demand and cost of production brought about by changed technology. In other words, modern agriculture is one which is highly dynamic and highly flexible and increasingly productive. The accomplishment of regional specialisation in agriculture would be possible by the identification of areas with varied productivity and growth potential. Thus, an analysis of spatial patterns of growth in agriculture becomes necessary in planning for the development of argiculture at the national and regional levels of different scale—macro, meso and micro.[1]

For the agricultural development of any country to have these characteristics, it must be commercial agriculture with the market and transportation facilities, that can provide efficient mobility of farm products.

Efficiency of Production

Efficiency of production involves technical and ecoonmic efficiency. The former deals with the physical relationship between input and output whereas, the latter deals with the cost-price relationship between input and output. In other words, we are interested in maximum efficiency at minimum cost.

The success of agriculture depends to a large extent on how successfully water requirement of various crops can be met. Availability of perennial irrigation encourages farmer to adopt more scientific techniques as well as intensive cultivation. It enables him to use the best quality of high yielding varieties of seeds at the right time to reap bigger margin of profits. The introduction of high yielding varieties has now changed the attitude of farmers in regions where they were accustomed to assured water supply either through rainfall or irrigation or both. Now the demand for irrigation facilities has risen. In fact, the Indian Agricultural Research Institute (IARI), New Delhi, conducted experiments on various crops in different regions which showed that critical periods for irrigation are first at the stage of grain root initiation, second, at the stage of flowering, and third, at the stage of grain formation. These results had the general validity that in areas where water availability is limited in rabi season, larger areas can benefit by extending irrigation facility over a wider area than intensive application in selected areas owned by relatively better off farmers.

In the case of both rice and wheat the I.A.R.I. experiments showed that if proper irrigation is not provided at the stage of root initiation, flowering and grain formation, yield might go down as much as 40 to 50 per cent. Thus, irrigation not only in right quantity but at the right time is also necessary. The planning of water resources has again to be on a regional basis taking into consideration the quality of the land, existing use and future capabilities.

Degree of Market Orientation

Degree of market orientation will determine the growth of agricultural production. Farmers who have not been previously accustomed to buying and selling must therefore acquire new skills and new attitudes with respect to farm operation and with respect to buying and selling if they are to increase production on their farms. Consequently, farmers in areas where the selling of agricultural products has been going on for some time can be expected to move into commercial farming more rapidly than those in areas previously characterised by only subsistence farming. To get a rough estimate of the present degree of market orientation of the farms of an area, one has to observe what commodities flow through the local markets. This can be done through a sample survey. Degree of market orientation helps in seeing place priorities among areas that have immediate high potential for agricultural development. It is only in these areas that present degree of market orientation can be estimated. Where a few farmers are already using the new technology, they can provide information with respect to yield responses, cost of additional inputs, local prices they are receiving for their products and so on.

In regions where transportation is readily available and market towns are already well developed, then existing prices may or may not be a dependable indices for use in establishing priorities among the immediate growth potential areas.

In India, since 1966, with the adoption of the strategy for securing higher agricultural production through the use of high yielding varieties of seeds more responsive to fertilizer use, there has been a market improvement in productivity, notably in wheat. It may be true that the new technology has not been adopted as widely and assertively as one would have desired. Around 1975, it can be said that change over to the new technology in wheat was fairly widespread and well established. However, in the case of rice and other foodgrains, the spread of the new technology is by no means complete and the results achieved so far have been very uneven all over the country. This is understandable because of the large area of varied landform, soil and climatic conditions under which rice is grown. There is also a sharp contrast in socio-economic and institutional factors prevalent between rice and wheat growing regions. Therefore, one of the arguments put forward is that the new technology has increased inequality within regions among farmers, the new technology being more accessible to the rich and better off farmers. But then the remedy lies in finding out ways and means of making way for a more even spread of new technology among different regions of the country; hence the case for rigorous regional analysis of agricultural development in India.

REFERENCE

1. Macro, regional analysis refers to the evaluation from a national angle; meso level to state and intrastate regions of the size of groups of Districts or Tehsils as the case may be while micro level refers to within District studies.

1

Objectives and Methodology

1.1 OBJECTIVES

THE MAIN objective of this study is to examine inter-regional variations in the cultivation of wheat and rice in selected regions of India between 1960-72. The new strategy under the title of green revolution, particularly in the case of cereals was implemented with success for the first time in the year 1966-67 which started with wheat first and was then extended to rice and other crops. Due to sharp contrasts in physio-agronomic conditions and the existence of socio-economic disparities among the people dependent on agriculture, the cropping pattern, productivity and agricultural practices vary widely between regions. This makes it very difficult to draw generalisations regarding the response of different crops to new technology in agriculture especially when the study is restricted to small regions.

Wheat and rice are taken up for this study because of their wider distribution in the country and the varying conditions under which the two crops are cultivated. These are also the principal crops in which the strategy for modernisation of agriculture has been in operation over the last 18-19 years.

Approach to the Study

In the first instance, a detailed study of the growth rate of agriculture and agricultural productivity has been undertaken in the selected regions of the country for the period 1960-72. Such a study brings out the major changes in crop productivity. The growth rate in different regions throws light on the varying performance of different regions and factors behind these variations.

A systematic analysis of variations in the porduction of the two crops has been undertaken by examining

i. Changes in cropping pattern of the two crops; and
ii. variations in the yields of the two crops in terms of factors having an important bearing on yields.

Changes in cropping pattern have been analysed in terms of variables such as, value of output per hectare, proportion of irrigated area of the crops, proportion of area under different crops and so on. Variations in yields of rice and wheat are analysed in terms of factors having an important bearing on yields. The new technology is relatively recent in origin and hence the patterns of change and the processes at work to bring about these changes have been studied with reference to the contrasts between traditional agricultural practices and the new technology within and between the selected regions. Under new technology are considered inpute like irrigation; agricultural practices which include modern implements, use of high yielding varieties, fertilizers, better manure, etc. These have been considered under factors that are controllable by the farmers. Among the uncontrollable factors affecting yields, cost-price relationship and size of holdings have been taken into account. Variations in soil conditions also affect yields. The different regions have been classified on the basis of differences in natural factors like soil types, land forms, rainfall, etc., and against this background regional variations (spatial and temporal) in yield of rice and wheat have been analysed.

This study is intended to throw light firstly, on the spatial concentration of rice and wheat in different regions, classified on the basis of differences in natural factors. This will bring out the inter-regional variations in the behaviour of different variables, influencing area under the crop, within and between regions.

Secondly, variations in the responses of rice and wheat to modern inputs within and between regions have been explained by a comparative analysis of rice and wheat.

Finally, the inter-regional variations have been examined vertically for different sizes of holdings. The holdings have been classified into three categories, namely, high, medium and low. Area and yield patterns, cost-price relationship, and application of inputs are studied for these three categories first, region by region, and then for all rice and wheat regions. This throws light on the yield patterns existing in different holding-sizes and could answer questions such as, is it true to say that in low yield regions, small and medium farmers predominate and vice versa.

This was analysed on the basis of primary data collected through a sample survey of 198 land holdings.

1.2 DESIGN OF THE STUDY

State Level Analysis

As a background to this study and to provide a spatial framework of an analysis, variations in agricultural productivity and growth rate of agriculture have been analysed between 1960-72 for seventeen major States covering twenty three principal crops. Such a study brings out the relative importance of different states in the cultivation of various crops; and the relative importance of different crops within and between the States.

The relative importance of different States in respect of cultivation of crops is given by the percentage of area under each crop in the State to area under the crop in the country as a whole. The relative importance of each crop in the State is given by the percentage of area under each crop in the State to total cropped area of the State.

Inter-Regional Analysis

State level analysis is followed by a district-wise analysis of variations in area and yield of rice and wheat for the period 1960-72. Firstly, area concentration (percentage of area under rice/wheat to gross cropped area in the district) of rice and wheat has been analysed for the 298 districts comprising the Seventeen States under study. For this purpose, the period

1960-72 has been divided into four time blocks, namely, average for 1960-63 (Time Block I), 1963-66 (Time Block II), 1966-68 (Time Block III), and 1970-72 (Time Block IV). The relative position of districts has been illustrated with the help of maps and tables at four points of time. Secondly, inter-regional variations in yield of rice and wheat have also been mapped to understand the variations—Spatial and temporal for the same period.[1] The analysis of maps throws light on inter-regional variations in the cultivation of wheat and rice.

Growth rate in area

Growth rates in area in rice and wheat between 1960-63 and 1970-72, between 1963-66 and 1970-72 and between 1960-63 and 1963-66 have been analysed with the help of step-wise regression in terms of the following explanatory variables :

i. Growth rate in gross cropped area;
ii. Percentage of area irrigated of the crop to area under the crop;
iii. Gross area irrigated as a percentage of gross cropped area;
iv. Growth rate in yield per hectare (in kilograms); and
v. Growth rate in yield per hectare (in rupees).

Inter-regional analysis of variations in area of rice and wheat covers those districts which have 10 per cent or more of the gross cropped area under rice and wheat. These are considered as core regions of rice and wheat.[2]

Growth rate in productivity

Growth rate in yield of rice and wheat have been analysed at two points of time 1967-68 and 1970-71.[3]

For the core regions of rice and wheat step-wise regression analysis has been used to explain yield of rice and wheat in terms of the following variables :

i. Fertilizer consumption per hectare;
ii. Tractors per thousand hectares;

iii. Tubewells and pumpsets per thousand hectares;
iv. Proportion of area irrigated of the crop;
v. Proportion of area under high yielding variety; and
vi. Price of the crop in rupees per kilogram.

1.3 CASE STUDIES

From the core regions of rice and wheat, seven rice villages and four wheat villages were selected. The criterion adopted in selecting them was that the district from which a particular village was chosen should represent either highest area concentration of rice or wheat in the State and/or highest average area under the crop concerned (rice or wheat) in the State. The criterion has been adopted uniformly except in the case of Ranchi in Bihar and Mathura in U.P.

1.4 SAMPLE SURVEY OF LAND HOLDINGS

Procedure for sampling

This is a method of sample selection where a unit in the universe is selected according to its relative "population" with respect to the "total population" of all the units in the universe. In other words, the probability of selecting a unit will vary directly with the "population" of the unit. Sampling was done at two stages, initially by villages and then the landholdings within each village.

Theory

Let the n units be denoted by $X_1, X_2 \ldots . X_n$ with the corresponding population $P_1, P_2 \ldots . P_n$. The population of all the units in the Universe will then be $P_1+P_2+ \ldots . P_n$ denoted by P i.e., $(P_1+P_2 \ldots . P_n=P)$.

Therefore, the probability of selecting the K_{th} unit will be $\frac{PK}{P}$ (i.e. the population of the K_{th} unit $\div$ total population of all the units . . . (1).

Practical application

The method adopted for selection should be such that condition (1) is satisfied.

e.g. Let N villages with population be given as below :

Sl. No.	*Population*	*Cumulative Population*
1	P_1	P_1
2	P_2	P_1+P_2
3	P_3	$P_1+P_2+P_3$
K	P_k	$P_1+P_2+ \ldots P_k-1+P_k$
N	P_n	$P_1+P_2 \ldots P_n-1+P_n$

Suppose we have to select four villages, (1) P/4 is first calculated. Let the integral part of it be 4 digited (say) and be denoted by I.

(2) A four digited number less then or equal to I is selected and let it be denoted by R_1.
(3) Second, third and fourth random number selected will then by R_1+I, R_1+2I and R_1+3I.

Hence, the four random numbers selected are :

(1) R_1, (2) $_1R+I$, (3) R_1+2I, and (4) R_1+3I

Procedure adopted for the selection of villages is as follows :

(a) Now, if the selected random number is less than or equal to P, the lst village is selected.
(b) If the selected random number lies between the two cumulative totals, the village corresponding to the largest of the cumulative total is selected.
(c) If the selected random number equals one of the cumulative totals the corresponding village is selected.

Theoretical justification

(1) The probability of selecting a given village (say No. 3) will be equal to the probability of the random number lying in the interval (P_1+P_2) and $(P_1+P_2+P_3)$, the magnitude thus depends on the width of the interval i.e. P_3. Similarly, the probability of selecting the K_{th} village would depend on the width of the interval $(P_1+P_2 \dots . P_k-1+P_k)$ i.e., P_k. In other words, the probability of selecting a village varies directly with the population of the village.

1.5 DATA BASE

Secondary data from published sources such as, Season and Crop Report estimates, Agricultural situation in India, Statistical abstracts of States, Bulletins of Rice and Wheat Statistics in India, etc., all published by the Ministry of Agriculture, Government of India. Data pertaining to fertilizers, was collected from Fertilizer Statistics, published by the Fertilizer Association of India. Apart from these main sources, others were also used for obtaining data on relevant matter. These are listed in the general Bibliography.

Primary data was collected on the field with the help of detailed questionnaire. In each village on the basis of purposive sampling, eighteen households were interviewed. The reference period was 1975-76. Households with a gross income of around Rs. 6,000 per annum were classified as high income households and those with gross income between Rs. 3,600-Rs. 6,000 per annum were classified as medium income households and those with a gross income below Rs. 3,600 per annum were classified as low income households.

1.6 TECHNIQUES OF ANALYSIS

The quantitative and cartographic techniques have been used at different stages of analysis. These are elaborated below :

Stepwise regression analysis

Multiple linear regression was used to explain the influence of certain explanatory variables[4] on,

i. Growth rate in area of Rice and Wheat at three points of time (1960-63 to 1963-66, 1963-66 to 1970-72, 1960-63 to 1970-72), first region by region and then for all the regions together.
ii Yield per hectare at two points of time (1967-68, 1970-72).

Factor analysis and clusters

In order to analyse the primary data pertaining to twenty four variables for about two hundred land holdings for rice and wheat factor analysis was used. First a multiple correlation matrix was prepared. Using factor analysis as the basis, highest factor leadings on each of the factors was worked out for large, medium and small samples of holdings. Variables were then selected to represent each factor keeping in mind the highest factor loading on each of the factors. In order to study the nearness of association between different land holdings on the basis of the variables chosen to represent each factor, cluster analysis was used.

Cartographic technique

Frequency distribution of different variables was analysed for selection of class intervals for mapping. The technique employed was simple choropleth mapping against the background of important isohyets. For the delineation of physical regions and sub-regions a composite map of various natural factors—landform, soils, climate—regional and seasonal variations provided a basis by which Districts were grouped according to homogeneity in these characteristics.

REFERENCES

1. Refer Chapter 4 for details.
2. The concept of core and periphery has physical and economic connections. Preliminary analysis of the concentration of these crops in the background grown of natural factors has been taken into account.
3. The analysis is attempted at only two points of time as it was difficult to obtain district-wise data on variables influencing yields for other years. In fact, even for these two years, for certain districts in the case of few variables State figures had to be substituted.
4. The explanatory variables chosen to explain growth rate in area and yield of rice and wheat are elaborated in Chapter 5.

2

Problems and Progress of Modern Agriculture in India

A Review

2.1 INTRODUCTION

AN ATTEMPT is made in this chapter to review the progress of agriculture since the beginning of the era of planning as it provides the background for analysing the factors that underlie the problem of minimising intra-regional and inter-regional variations in agricultural productivity and cropping pattern.

During the period 1951-54 to 1958-61, total crop output in India is estimated to have increased by 27.8 per cent, a compound annual rate of 3.57 per cent.[1] In Punjab, Tamil Nadu, Gujarat, Karnataka, Rajasthan, Kerala and Madhya Pradesh rate of growth was higher than the national average. West Bengal had the lowest rate of increase followed by Orissa. The study by Minhas and Vaidyanathan also revealed that variation in growth rates was even more marked at the district level. Of the 268 districts studied, 67 achieved an annual increase of 7.5 per cent or more, 114 districts covered an area of 122 million acres and had a growth rate of above 5 per cent per year. Fifty Nine districts recorded growth rate between 2.5 per cent and 5 per cent per year. This works out to about 65 per cent of the districts recording growth rates in crop production higher than the average rate of growth of population in India.

In other 95 districts, annual increase in output was less than 2.5 per cent out of which in 24 districts crop production actually declined during this period. According to Minhas and Vaidyanathan, a major part of the this output increase came about due to increase in growth of gross crop area alone (in Punjab and Rajasthan). In Karnataka and Madhya Pradesh, the contribution of area (1.63 per cent and 1.65 per cent) has been fairly significant. However, in West Bengal and Assam this was not true. In Kerala, Tamil Nadu, Karnataka, Punjab and Madhya Pradesh, the contribution to crop production through yield increase was higher than the national average. In Bihar, though the contribution to production by yield increase was low, yet good production was partly the result of good yields. In Rajasthan and West Bengal yields had actually declined.

At the national level between 1951-54 and 1958-61, crop pattern changes did not affect the growth of crop output significantly; its contribution to crop output growth was only 0.29 per cent. The proportion of area covered by cereals went down between 1951-54 and 1958-61. There was a slight fall in the share of rice, from 24.35 per cent to 23.79 per cent. The share of wheat increased from 7.99 per cent to 9.22 per cent.

During 1960-63 to 1967-70, area under rice as a percentage of gross cropped area remained almost the same and that of wheat increased by 1.16 per cent. But, between 1967-70 and 1970-72, the area under rice to gross cropped area increased from 22.88 per cent to 24.70 per cent. The share of wheat increased from 9.22 per cent in 1967-70 to 11.35 per cent in 1970-72.

However, in few States like Gujarat, Tamil Nadu, Punjab and Andhra Pradesh, changes in crop pattern did contribute to output growth.

2.2 COMPONENTS OF THE GROWTH OF CROP OUTPUT

Minhas was the first to decompose the production series and measure the influence of the individual components at the All-India level. After Minhas, important contributions came from Ashok Parikh who studied all the States individually[2] and V. Venkateswarlu who confined his study to Andhra Pradesh.[3] Later, Shetty made a similar study at the All-India level as well as at the regional levels for the period 1931-65 for All-India and 1925-55 for regional.[4] All these studies point to one common conclusion, namely, even for the period 1920-21 to 1964-65, the major contribution of growth in crop production came from an expansion of area and changes in crop pattern. These taken together account for four-fifths of total growth. Even Punjab was not an exception to this as far as the first decade of planning is concerned.[5]

Many other studies have been attempted at the State and below State level. For the period 1949-69, Mishra carried out the analysis for the districts of Gujarat and studied the interaction between area and yield, yield and crop pattern, area and crop pattern and also area, yield and crop pattern all taken together. The results of the study showed that between 1949-59, contribution of area expansion to production increase was important in seven out of sixteen districts. Between 1959-69, no significant contribution was made by area expansion in any of the districts.[6]

Between 1960-63 and 1967-70, there was a shift in proportion of area devoted to different crops. But the shift was not substantial at the national level. In States, like Orissa, area under all crops had increased. This could be due to increase in total cropped area and extension of irrigation. There has been hardly any change in the proportion of area devoted to food crops at the national level. However, for the individual States, there are variations. These Inter-State variations are elaborated in Chapter 3.

2.3 INTER-REGIONAL VARIATIONS IN AGRICULTURAL PRODUCTIVITY

There are a number of studies concerning inter-regional variations in agricultural productivity. For example, S.B. Tambad studied inter-district variations in Karnataka State. Similar researches were made for Andhra Pradesh by Gopalakrishnan and Ramachandra Rao.[7]

Some reasarch workers investigated into whether the size of holding has any influence on the selection of cropping pattern. Abdul Majid investigated the problem for U.P. for the period 1959-60 to 1961-72. He observed direct influence of the size of holding on the pattern of crops in which small farmers devoted most of the land to food crops.[8]

In most of these studies the technique of coefficient of

correlation was employed to measure the impact of several explanatory variables on yield. All the studies revealed the absence of a positive trend in per acre yield in almost all the important crops, for the period upto 1955-56.

Between 1960-63 and 1967-70, growth rates in yield of most crops do not indicate a positive trend at the national level. However, in the inter-State analysis, few crops do indicate a positive trend in yield growth rates in few States.[9]

2.4 USE OF MODERN INPUTS

Fertilizer use

Three distinct areas have been covered so far in the studies on the use of fertilizers.

i. Experiments have been undertaken on the basis of available data to determine the optimum dose for different types of crops in various regions of the country with a view to maximising profit to the producers.
ii. Some studies were carried out on the basis of data collected from the actual use of fertilizers by the farmers in their fields. Attempts were made to investigate into the demand for fertilizers at different price levels.
iii. Studies relating to the use of fertilizers have also gone into the existence of complementarity between the use of fertilizers and irrigation water. But the conclusions of these studies differ.

Abraham and Rakheja fitted the production function using only irrigation and fertilizer data to represent inputs and production data for wheat and rice for the years 1951-52 to 1964-65. The exercise was carried out at the national level followed by a study of rice in the eastern region and wheat in the western and northern region.[10] Fertilizer use was significant in the case of rice and for wheat irrigation was significant.[11] Two leading institutions, National Council of Applied Economic Research (NCAER) and Programme Evaluation Organisation (PEO) carried out an all-India study to analyse factors that influence the use of fertilizers. These studies observed that size of holding, irrigation, type of tenancy influenced the use of fertilizers. Lack of adequate credit restricted the consumption of fertilizer.[12]

Present study has made a systematic attempt to measure the influence of consumption of fertilizers per hectare on wheat and rice yields for 1967-68 and 1970-71 alongwith other explanatory variables such as price, proportion of high yielding variety, irrigation, tractors per thousand hectares.[13]

Use of high yielding variety seeds

The technological change introduced recently is referred to as green revolution. The high yielding variety seeds have been introduced particularly in the cultivation of wheat and rice since 1965. The core of new technology is the high yielding variety seed. Minhas and Srinivasan worked on the use of high yielding varieties.[14] In fact, extensive work has been done on the evaluation of the impact of high yielding variety on the yield of crops.

The major conclusions of these research workers can be summarised as follows :

i. With the adoption of high yielding variety, although the farmer's total cost of cultivation would increase several fold, yet, the increase in gross income from the produce would be much larger, leaving a substantial profit to the producer.
ii. The major share in the increase in the cost is due to fertilizer. But expenditure on wages would also increase.
iii. Small farmers would lag behind in the adoption of the new technology.

However, some studies have also proved that the small farmers might not lag that far behind the others in the adoption of high yielding variety. Muttiah observed this in the context of his study on Green Revolution in Madras.[15]

2.5 PROGRESS OF MODERN AGRICULTURE IN INDIA

The Indian planners have been concerned with the basic problem of increasing agricultural production. Although good progress has been achieved, yet, the agricultural sector continues to be backward for the following reasons :

i. Infrastructure facilities such as, assured irrigation, roads, power, market are inadequate.
ii. The irrigation potentials already created are not fully utilised due to lack of, on the farm works.
iii. Low productivity of crops due to failure of monsoon, floods, institutional factors and socio-economic conditions of farmers.

Although, area under assured irrigation has been increasing, proportion of the net irrigated area to the area sown, at the All-India level was less than 20 per cent. In States like Punjab, it is as high as 70 per cent, while in Bihar even after the provision of irrigation facility through canals or tubewells, the actual utilisation has not been more than 50 per cent.

The first step for intensifying cultivation was taken in 1959 with the introduction of Intensive Agricultural Development Programme (IADP) in seven districts of India.[16] The IADP approach envisaged adoption of a package of practices by participating farmers in terms of fertilizers, seeds, pesticides and use of improved implements for cultivation and inter-cultural operations. The use of organic and green manures was also recommended. The machinery was strengthened. An organised effort of this type resulted in an increase in the use of nitrogenous fertilizers from 56,000 tonnes in 1960-61 to 2,40,000 tonnes in 1965-66 in these seven districts. Similar increases were recorded in the use of pesticides, improved implements and disbursement of agricultural credit. In Ludhiana district (Punjab) results were far more remarkable than in Shahbad district of Bihar. Constraints in provision of irrigation facilities to participating farmers in Shahbad were not conducive to optimality in productivity specially of high yielding variety of seeds. The main effort of integrated agricultural development programme should have been to overcome these difficulties, which still are an impediment to increase in agricultural productivity. In 1971, a task force was set up under the Chairmanship of Minhas, which in its interim report had highlighted the problems and suggested measures to overcome the difficulties especially in canal irrigated areas where huge investments in the river valley projects have not led to the desired results. There should be a better coordination of various programmes of agriculture in command areas of major irrigation projects including the layout of field channels for taking irrigation water to the fields. Thus a beginning was made to evolve an integrated approach for the development of agriculture in these areas. It must, however be mentioned that in the context of a spatial strategy of agricultural development these efforts should be viewed as *ad hoc* approach in limited areas.

High yielding variety of seed brightened hopes for a break through in agricultural production. Such a variety of seed loses its potency quickly over the years and it has to be replaced by fresh high potency seeds. For example, investigations made in high yielding variety of wheat seed in rabi season of 1972 revealed that lack of availability of good quality seed was responsible for low productivity. There was an absence of proper seed certification agency and lack of technical supervision over lands devoted to the production of improved seed. Also the programme of high yielding varieties is likely to suffer a set back, if plant protection measures are not taken at the proper time during the growth phase of crops. The investments being higher, the farmer cannot afford losses due to crop failures.

The new dwarf varieties of rice and wheat evolved by agricultural research stations provided a break-through in raising yield per hectare. Fertilizer requirements of these high yielding varieties of seed were found to be as high as 100 kgs of nitrogen per hectare. These seeds offered the possibilities for a break-through in areas of assured water supply. Since 1966, considerable progress has been made in Punjab and Haryana due to the adoption of high yielding variety covering large areas of cultivation irrespective of differences in the size of holdings.

2.6 HIGH YIELDING VARIETY

Rice

In Kharif season of 1968, the average yield for 704 harvests of high yielding varieties of paddy reported by selected participants worked to 39.9 quintals per hectare. The average yields of the high yielding varieties of paddy are given below in the table.[17]

HYV of Paddy	*Average Yields* (Quintals per Hectare)
i. Taichung Native I	29.50
ii. IR-8	42.90
iii. D.A.+27	37.91
iv. Taiwan-3	24.54
v. Others	43.44
vi. All varieties	39.91

Higher yields were also due to relatively better climatic conditions in 1968. The prevalence of pests was also reported to be less.

Punjab reported highest yield after Jammu and Kashmir and Karnataka. Among the various high yielding varieties, the largest number of harvests were for IR-8 (270) and A.D.T. (128). Taiwan-3, quite popular some time ago has been replaced by other varieties.

The dwarf varieties of paddy that became popular in India, namely *Cauvery*, *Ratnas Jaya*, *Vijaya* (IR-20), *Pankaj* and *Jagannath* mature at the same time as the traditional varieties.

The process of evolving new varieties continues as the physical factors that contribute to the success of new varieties vary from region to region. However, before these varieties are popularised in any area, it is important to find out the duration of the varieties presently grown, staggering of planting within the season, prevalence of pests and diseases. The variety recommended should have a duration long enough to exploit the full growing season and short enough to insure against crop failure due to excessive cold or drought at the time of ripening.

2.7 NATIONAL PROGRAMME TO STEP-UP RICE PRODUCTION—A REVIEW OF THE PROGRESS MADE IN EVOLVING IMPROVED VARIETIES OF RICE TO SUIT REGIONAL CONDITIONS

The Indian Council of Agricultural Research (ICAR), prepared a nationwide programme to step-up rice production.[18] The programme took note of variations in agro-climatic conditions.

Rice is grown under varied agro-climatic conditions from water logged areas to upland regions and hence, appropriate high yielding variety should be selected. In predominant rice growing regions, the use of high yielding variety is rather sporadic and needs extensive adoption over the entire region. For this a massive programme of the HYV is required.

On the basis of facts collected by UNI, (1975), following are the highlights of the high yielding varieties programme being carried out in different parts of the country. Improved cultural practices like close spacing, early weed control, late top dressing of nitrogenous fertilizers and good water management are also important. According to the All-India Coordinated Rice Improvement Project (AICRIP) at Hyderabad, India has already made good progress in rice research. The AICRIP, coordinates the work of 100 institutions. Now there are so many new varieties in contrast to 1965 when only Taichung Native-I was introduced. A particular region can now choose a variety depending on the length of the available growing season and also the grain type it prefers to produce, where pests and diseases pose a problem, dwarf varieties can be used. The management display trials conducted promise a 60 per cent increase in yield by improving only agricultural practices. Other new varieties developed by the AICRIP include RPW-6-12, RPW 6-13, RPW-6-17 and *Kakatiya* which were resistant to gallmidya, a problem in most Asian countries, and suit Indian conditions.

The Tamil Nadu Agricultural Department released ADT-30 which is a cross between IR-262 and ADT-27 and has a duration of 80-90 days from seed to seed. The long bold grain is good in quality with 11.25 per cent proteins. The variety is recommended for Kuruvai season i.e. June-July season and also Navarai or summer season in all the areas. It has an average potential yield of 5,261 kg/ha according to the Tamil Nadu Agricultural Department.

Another new variety is ADT-31, a derivative of a cross between IR-8 and CU-340 evolved at Aduthurai Centre. It is a semi-dwarf plant type. It matures in 100-104 days. Its potential yield is 6,198 kg/ha. It has performed better than *Karikalan* in Thanjavur district. *Bhavani* is another variety released by the department. It is better than *IR-8* and Taichung native and has a potential yield of 6,8Ç5 kg/ha.

Another variety of paddy called *Pusa-33* (a scented variety) is now under trial at the Rice Research Station at Aduthurai. It is a cross of *BC-1163* and *Ratna*. It has been tried on 5 acres as an experiment and the results were found to be satisfactory. *Vaigai* is another variety developed at the paddy breeding station of Tamil Nadu State Agriculture University. This variety matures in 115-120 days after owing. It grows to a height of 3 feet and requires moderate doses of fertilizers. This is recommended for most of the southern districts and can be grown in the three seasons.

In Kerala, the Pattambi Rice Research Station has evolved three new rice varieties with red rice. They are *Culture 1285*, *6473* and *Culture 12035*. These were produced by crossing a hybrid evolved by crossing *Annapoorna* and *IR-8* with *IR-8* again. It gives more yield than IR-8 and has a duration of 130-135 days. *Culture 6473* is a short duration variety and gives yield upto seven tons per hectare in fields. Culture 1265 is a cross of *P.T. DGM-10* with IR-8 and has a duration of 100-105 days.

In West Bengal, at Chinsurah three high yielding varieties have been introduced by cross breeding *Dee-Gee-Woo-Gen*, (Taiwan) with NC-1626, a local variety from Burdwan. This variety had given yield equal to or higher than *Jaya* and *IR-8*. They are highly resistant to disease.

In Punjab, dwarf varieties such as IR-8 and Jaya have yielded more than 10 tonnes of grain per hectare at several places. New dwarf varieties are under test. A new variety named *Palman-579* is capable of yielding 70 quintals of grain per hectare in shorter period than IR-8 or Jaya. Two other varieties are *RP-5-3* and *Hybrid Mutant-95*. RP5-3 is a short statured variety. It gives almost the same yield as Jaya but it is more resistant to pests. Another variety evolved by the Punjab Agricultural University is M-95 which matures quickly and is good for late sowing. It ripens seventy days after transplantation. Its plants are short and attain a height shorter than Jaya and local varieties. It is better than Jaya and other local varieties in the sense that the latter, if planted in second half of July, give very low yield and sometimes can fail completely due to early onset of cold weather. Jaya and IR-8 if sown early are found to give higher yield than M-95.

A trial conducted at Rajamundari district in Andhra Pradesh by the Central Tobacco Research Institute has shown that farmers can earn an additional income of Rs. 250 per hectare if paddy is grown in strips during the Kharif season followed by virginia tobacco.

The International Rice Research at Manila has released IR-26 which is better than IR-8. It is insensitive to changes in day length and reached maturity in 126 days. It has a stronger stem than IR-20 and therefore a better yield. It suits both irrigated and rainfed low land conditions.

The farmers in Trivandrum, Palghat and Coimbatore area have reported record yield. However, merely an introduction of HYV is not sufficient. The farmers have to adhere to the technological norms pertaining to water, agricultural practices and so on, in order to obtain best results.

2.8 HIGH YIELDING VARIETY

Wheat

The high yielding variety of wheat introduced and tried so far are, *Mexican Variety*, *Luna*, *Rojo* and *Sonara 64*. Here again, the basic element in the new technology is the dwarf nature of the exotic varieties enabling the application of a high dose of fertilizer with the consequent gain of a much higher yield compared to the hitherto available varieties of the crop. Between rabi season of 1965-66 and 1967-68, few other varieties have been added. These include, *S-227*, *S-308* and *P.U.-18*

(Mexican in origin) and K-68 indigenous higher yielder evolved in U.P. Thc release of PU-18, S-227 and S-308, for general cultivation during rabi 1967-68 somewhat eclipsed earlier varieties of wheat of Mexican origin, as these had shown better performance in yields, besides their better grain quality. Then came the triple dwarf variety. All these varieties were adopted by cultivators in the States of Punjab, Rajasthan, Haryana and Uttar Pradesh.

The indigenous varieties of wheat are sown during October and early November. The dwarf varieties have a short duration and therefore delayed sowing would prove a more beneficial crop. Higher yields were reported for PU-18, followed by S-227 and Luna-Rojo varieties and the least responsive was K-68. Punjab and Haryana complained about inadequate supply of seed. Problem of obtaining credit and acute shortage of labour was also reported from these States. The proportion of cultivators wanting to continue the use of high yielding variety was 95 per cent.

2.9 National programme to step-up wheat production

The ICAR prepared a national programme to step-up wheat production in States by recommending the adoption of high yielding varieties of wheat. Following are the highlights of the nationwide programme carried out by the ICAR.

The exact nitrogen requirement of dwarf wheat will depend on the crop preceding. If wheat follows maize or pulses the requirement of nitrogen dose decreases. If wheat follows Jowar or paddy there is need for more direct application of nitrogen, as little nitrogen is left. In fact pulse-wheat rotation showed highest net income followed by paddy-wheat rotation. Jowar-wheat rotation gave the lowest income.

Pulse-wheat rotation economises on the use of fertilizers as a smaller dosage of nitrogen is required in the next season. These conclusions were drawn after experiments at U.P. Institute of Agricultural Sciences, Kanpur, for 1967-68 and 1968-69 and 1969-70, on nitrogen requirement of dwarf wheat grown after high yielding kharif crops such as maize, Jowar, paddy and urad.

The Institute also conducted a study to show that wheat

responds significantly to nitrogen application. The application of 60 kg of nitrogen per hectare gave the highest yield.

Among the new wheat varieties, in Punjab, *Kalyan Sona 227*, is at present most important like PU-18. It is a two gene dwarf and can stand heavy doses of fertilizers. It has a yield potential of over 6000 kg per hectare. A new variety *WG-357* which combines in it high yield potential of the semi-dwarf variety PU-18 and good grain quality of *C-273* was also released for cultivation by the Punjab Agricultural University. The grains of WG-357 are big and possess good food qualities. Another two-gene dwarf wheat variety, *WG-377* developed from the cross between *WG-143* and *USA-255* and Kalyan Sona has been released for cultivation which yields eight times more than Kalyan Sona. In Punjab efforts are being made to introduce varieties which are more resistant to diseases and which can take heavier doses of fertilizers. The three-gene dwarfs which have shorter plants than the two-gene dwarfs are also under test. *WL-212* is at an advanced stage of testing. It out-yielded Kalyan Sona-227 by a margin of 10 per cent. A number of high yielding varieties of wheat that will increase yield three times in hill areas have been evolved by the Vivekananda Agricultural Research Laboratory at Almora, Uttar Pradesh. For hilly areas cultivation of rust resistant varieties of wheat such as Kalyan Sona and PU-18 should be replaced by "Girija" a new high yielding wheat variety developed by Simla Wheat Research Station for cultivation in these areas. This new variety possesses a high degree of resistance to rust and other diseases and it was developed by a process of high hybridisation. It gives an average yield of 1250 kg per hectare under rainfed and irrigated conditions. The yield is found to be higher than that of Kalyan Sona.

In Gujarat, the Wheat Research Station in Junagadh has evolved nearly 2,000 varieties, both from India and abroad, varieties received from Mexico have been put under trial. There are 94 varieties from different States. Twenty four of the varieties evolved are included under the All-India Coordinated Trials.

The *NP-824* and Kalyan Sona varieties under cultivation have been replaced by Junagadh 1-7 in irrigated areas. This has turned out to be a high yielding variety in Gujarat particularly suited to relatively dry areas of Gujarat. A new variety A-206 has also been released for cultivation. Twenty nine wheat varieties suited to different agro-climatic conditions of the country have been identified by wheat researchers. The new varieties have been developed by the Agricultural Universities and State Departments of Agriculture of Punjab, Uttar Pradesh, Rajasthan, Madhya Pradesh, Gujarat, Maharashtra and the Indian Agricultural Research Institute, New Delhi.

For rabi wheat, *Sonalika*, the second best high yielding variety has been recommended by ICAR for cultivation. This should be grown in the latter half of November for best results. Even when planting is done in December it has proved to be a good variety due to late harvest of paddy, potato and other crops.

2.10 FERTILIZER RESPONSES

Nitrogen (N), Phospherous (P) and Potassium (K) play an important role in crop development.

In order to recommend fertilizer application on a scientific basis, field experiments have been conducted on many places over the last fifty years. The results have generally shown that in India, almost all our soils need nitrogen application, 85 per cent need additional potash to increase crop yield by 50 to 100 per cent. It was also observed that the interaction of three types of fertilizers resulted in a significant rise in yield. The economic use of fertilizer depends upon :

i. availability of adequate irrigation or rainfall;
ii. the prices which the farmer pays to get them and prices which he gets for the product; and
iii. the extension of knowledge among farmers about the use of fertilizers.

The important factors which influence the consumption of fertilizers are :

i. changes in the total crop area under cultivation including that for multiple cropping ;
ii. changes in area under irrigation ;

iii. area under high yielding varieties ; and
iv. cropping pattern.

The studies made by Programme Evaluation Organisation showed that once the farmer uses chemical fertilizer, he increases its consumption in subsequent years.

Fertilizer and high yielding varieties—Rice and Wheat

The high yielding varieties of paddy and wheat are well known for their high response to fertilizers. On the basis of the results obtained in the coordinated projects of the ICAR, the optimum dose of nitrogen for these varieties are as follows :

Rice

One tonne of nitrogen applied at 100 to 120 kg per hectare will give additional yield of 15 tonnes per hectare.

Wheat

One tonne of nitrogen applied at 100 to 120 kg per hectare will give an additional yield of 20 tonnes per hectare.

The NCAER investigated the factors influencing the use of fertilizers by the farmers.[19] The study was based on large sample of 4,000 households selected from all over the country. It covered major crops, notably rice, wheat, maize, jowar, Sugarcane and cotton. The survey covered 1968-69, 1969-70 and 1970-71. The major factors analysed in this study were size of holdings, the extent of irrigation facilities available with farmers, cultivation of HYV crops vis-a-vis traditional varieties; availability of credit to the farmer, characteristics of the household, such as age and education level of the head of the household and tenurial status.

This study revealed that irrigation, extent of high yielding varieties of crops cultivated and credit availability had a dominant effect on fertilizer use. The small farmers seem to have applied fertilizer on a smaller proportion of cultivated area but more intensively compared to bigger farmers. On the other hand, farmers having larger holdings perferred to apply fertilizer in small quantities over a large proportion of the cultivated area.

The NCAER estimated for the first time cropwise fertilizer consumption for 1970-71 based on reported use levels of fertilizer for the major crops. These estimates showed that about two-third of the total fertilizer input was for food-grain crops while about one-third was for other crops. Rice and wheat alone accounted for a little over 50 per cent of the total fertilizer used in 1970-71. The most important factor that induced the farmers in their decision to use fertilizer was the expectation of higher yields as a result of the adoption of high yielding varieties. The motivation came from Block Development Officers, relatives and other farmers who had benefited from the use of fertilizer. The major findings of the NCAER study, to the extent they are relevant for this analysis are summarised below.

Rice

The cultivators of rice used on an average 34.79 kg of N, 9,60 kg of P_2O_5 and 3.14 kg of K_2O per hectare in 1968-69. In 1970-71 they used on an average 39.82 kg N, 8.21 kg of P_2O_5 and 1.62 kg of K_2O per hectare. Although, this is hardly an increase, the proportion of gross cropped area on which the fertilizers have been used has shown an increasing trend. It rose from 25.46 per cent of gross cropped area in 1968-69 to 46.98 per cent in 1970-71. Fertilizer input was substantially higher in irrigated areas under rice than those in unirrigated areas. For example, during 1970-71 the rate of application of nitrogenous fertilizers per hectare in irrigated rice lands was estimated to be 48.41 kg/ha, whereas in unirrigated areas, it was 29.84 kg/hectare.

Irrigated Rice

In the case of HYV of rice grown on irrigated land, the cultivators applied on an average 67.97 kg of N, 11.52 kg of P_2O_5 and 2.3 kg of K_2O per hectare (1970-71). This is an increase over the two earlier years. Thus, there was an increasing trend in the use of nitrogenous fertillzers per hectare on HYV's of rice in irrigated areas during 1968-69, and 1970-71. So far as traditional varieties of rice on irrigated lands were concerned, there appeared to be no significant increase in fertilizer use (52.32 per cent in 1968-69, 53.54 per cent in 1969-70 and 55.44 per cent in 1970-71).

Unirrigated-Rice

For cultivation of HY variety of rice on unirrigated land, the farmers had used on an average 31.82 kg/ha of N, 19.52 kg of P_2O_5 and 1.07 kg/ha of K_2O in 1968-69, compared to the average level of use of 30.76 kg/ha of N, 5.73 kg/ha of P_2O_5 and 0.15 kg/ha of K_2O on traditional variety of rice. By 1970-71, the input of fertilizer on unirrigated lands growing modern varieties of paddy, had gone upto 44.29 kg/ha of N, 6.30 kg/ha of P_2O_5 and 2.36 kg/ha of K_2O. On traditional variety of rice, the use of N had gone down to 26.47 kg/ha. The input of P_2O_5 kg/ha had shown a slight increase.

Wheat

The cultivators of wheat used on an average 47.25 kg/ha of N, 18.02 kg of P_2O_5 and 2.2 kg of K_2O per hectare in 1968-69. The proportion of wheat area under inorganic fertilizers to gross cropped area under wheat for the same year was 37.42 per cent. In 1970-71, the cultivators used larger doses of fertilizers per hectare (50.66 kg/ha of N, 3.93 kg/ha of P_2O_5 and 1.65 kg/ha of K_2O). Nearly 49.33 per cent of the gross cropped area under wheat was covered by fertilizer inputs.

Irrigated-wheat

In the case of cultivation of wheat in irrigated lands, the farmers applied on an average 53.57 kg of N, 3.40 kg of P_2O_5 and 1.23 kg of K_2O per hectare in 1970-71. This does not show a clearly increasing trend over the earlier years; yet the proportion of gross cropped area on which the fertilizers have been used showed a consistently increasing trend from 58.00 per cent in 1968-69 to 68.71 per cent in 1970-71.

Unirrigated-wheat

In the case of unirrigated wheat, the cultivators used on an average 22.44 kg of N, 13.79 kg of P_2O_5 and 0.23 kg of K_2O in 1968-69. In 1970-71, they used an average 27.56 kg of N, 8.11 kg of P_2O_5 and 5.03 kgs of K_2O per hectare in 1970-71. The proportion of area under inorganic fertilizer inputs to gross cropped area was 15.75 per cent in 1970-71, slightly lower than in 1968-69.

For the cultivation of traditional varieties of wheat on unirrigated land, the farmers have used on an average 22.44 kg of N, 13.79 kg of P_2O_5 and 0.33 kg of K_2O per hectare on 17.11 per cent of gross cropped area in 1968-69. In 1970-71 they used 17.91 kg of N, 3.18 kg of P_2O_5 and 0.28 kg of K_2O per hectare on 13.37 per cent of gross cropped area.

However, the farmers using HY varieties of wheat applied fertilizers over larger area compared to the cultivators of traditional varieties of wheat on irrigated land. On an average, cultivators of HYV's of wheat on irrigated land used 59.03 kg of N, 3.49 kg of P_2O_5 and 0 87 kgs of K_2O per hectare on 88.48 per cent of gross cropped area. On traditional varieties of wheat, they applied on an average 42.46 kg of N, 3.22 kgs of P_2O_5 and 1.95 kg of K_2O per hectare on 47.25 per cent of gross cropped area. Similarly, it appears that there is a declining trend in the rate of fertilizer use on traditional varieties of wheat on unirrigated land in 1970-71 as compared to 1968-69. In contrast, in the case of cultivation of modern varieties of wheat on unirrigated land, the fertilizer use by the cultivator has shown an increasing trend in 1970-71.

Factors influencing the use of nitrogenous fertilizers

The above review brings out the fact that, in the case of rice, factors such as: (i) percentage of irrigated area to gross cropped area under rice, (ii) percentage of area under HYV of rice to total cultivated area under rice, (iii) income level of household, and (iv) percentage of area owned to the total cultivated area of the household have statistically significant and positive influence. In the case of Phosphatic fertilizers on rice, percentage of irrigated area to gross cropped area and percentage of owned land to total cultivated area were significant and had a positive influence.

In the case of wheat, factors such as : (i) percentage of area under high yielding varieties to gross cropped area under wheat, (ii) percentage of irrigated area to gross cropped area, (iii) income level of the cultivated household turned out to be statistically significant and had a positive influence. Although the credit factor has a positive influence on fertilizer use, its value turned out to be insignificant. This goes to show that credit availability for fertilizers, other factors, remaining the same, is not a major constraint on fertilizer use by the cultivators of wheat.

According to the survey, whatever may be the other motivating factors, the most important reason for fertilizer use was the expected increase in yields and output. It was also found that majority of cultivators started using fertilizers before 1968 and the proportion of users of fertilizers was on an average around 3 per cent.

Regional variations in fertilizer consumption

Fertilizer consumption and the increase in area under high yielding variety have varied widely from region to region. This is due to varying agro-climatic conditions, cropping pattern, availability of irrigation facilities and economic return from these high yielding varieties. These have had a considerable impact on the consumption of fertilizers and account for regionwise variations. In 1970-71, northern region exceeded its targets of coverage of HYV and multiple cropping programme. Consumption of fertilizers has been rising in all the five regions of India (South, West, North, East and Central). There was hardly any increase in southern region until 1970-71.

The northern region has the highest fertilizer consumption per unit of cropped area in 1971-72 and 1972-73. This was followed by southern region. Between 1969-70 and 1972-73, rate of increase in fertilizer consumption per unit of cropped area is highest for north (20.56). For the southern region, consumption of fertilizer per unit of cropped area increased by 3.47 kg/ha. For the West, it increased by 6.53 kg/ha In Central India between 1971-72 and 1972-73 consumption of fertilizers per unit of cropped area fell slightly.

2.11 CONCLUSION

It is common knowledge that the new farm strategy based on the use of high yielding varieties supported by adequate irrigation and timely application of fertilizers, fungicides and pesticides has had a better success on the wheat farms than on the rice farms.

However, over the last few years steps have been taken to correct the disparities in yield between wheat and rice as a result of which certain regions in the country like Punjab and Tamil Nadu have been reporting encouraging levels of output from year to year. But Bihar, Orissa, West Bengal, Uttar Pradesh and Madhya Pradesh, accounting for 65 per cent of rice growing area, have been reporting a stagnation in both production and productivity. Part of the cultivation problems in coastal or low lying areas has remained as acute as ever.

Though there has been a nation-wide programme for acceleration in the adoption of modern agricultural practices, there is a wide gap between the break-through achieved on sample demonstration farms and Performance in the farmers' own holdings. The inadequate involvement of the farmer was also responsible for this gap. The reluctance of the farmer arises from inequalities relating to the supply of inputs, which arrive too late and/or, are in insufficient quantities at high prices.

It is important to activate production in States like Orissa where initiative of the farmer is not aroused enough to introduce efficient cultivation, and in Bihar and Uttar Pradesh where ground water supplies should be harnessed better to solve the irrigation problems in those fertile tracts which are far away from river borne water supplies.

The foregoing review is an attempt in summarising the salient features of the new strategy for modernising agriculture both in terms of the results of studies in agriculture and achievements in laboratories some of which have reached the farmers and it is a continuous process because of the complexity of man environment relationship.

REFERENCES

1. B.S. Minhas and A. Vaidyanathan (1965), 'Growth of Crop Output in India, 1951-54 to 1958-61—An Analysis of Component Elements, *Journal of the Indian Society of Agricultural Statistics*. 17(2). pp. 230-252.
2. Ashok Parikh, (1970), 'Cropwise Districtwise Production Function,' *Indian Journal of Agricultural Economics*. 25(1) January-March.
3. V. Venkateswarlu (1965), 'Growth in Agricultural Output in Andhra Pradesh During the Period 1952-53 to 1961-62', *Arthaniti*, 8(2) July. pp. 157-170.
4. S.A. Shetty (1970), 'Longterm Trends in Farm Production in U.P. 1901—1961,' Ph.D. thesis, Lucknow University (Unpublished).
5. Indian Council of Social Science Research (1974), *A Survey of Research in Economics*—Vol. III, Agriculture (1).
6. U.N. Mishra (1971), 'Growth of Crop Output in Gujarat : A Component Analysis', *Anvesak*, I(1), June. pp. 1-15.
7. M.D. Gopalakrishnan and T. Ramakrishna Rao (1964), 'The Regional Variations and Agricultural Productivity in Andhra Pradesh'. *Indian Journal of Agricultural Economics*, 19(1). pp. 227-236.
8. Abdul Majid (1963), 'Crop Pattern and Size of Cultivated Holdings', *Indian Journal of Agricultural Economics* 18(1) January-March, pp. 24-31.
9. Refer Chapter 3 for details.
10. Abraham, T.P. and S.K. Rakheja (1977), 'An Analysis of Growth of Production of Rice and Wheat Crops in India,' *Indian Journal of Agricultural Economics*, 22(3), July-September, pp. 1-15.
11. It is only to be expected that rice in the eastern region is rainfed while wheat is grown with the help of irrigation.
12. (a) National Council of Applied Economic Research (NCAER), Factors affecting Fertilizer Consumption (Problems and Policies), New Delhi, (b) Programme Evaluation Organisation (PEO), Study on the use of Fertilizers and Agricultural Production, Planning Commission, Government of India, New Delhi (1967).
13. For Details refer Chapter 5.
14. B.S.Minhas and T.N. Srinivasan, 'New Agricultural Strategy Analysed', *Yojana* 10(1), January, 1966, pp. 20-24.
15. C. Muttiah (1966), 'The Green Revolution—Participation by Small and Large Farmers'. *Indian Journal of Agricultural Economics*, 26(1), January, pp. 53-66.

16. R.C. Arora (1976), Development of Agriculture and Allied Sectors. An integrated area approach.
17. Indian Council of Agricultural Research (1971-72)—*All-India Coordinated Agronomic Experiments Scheme Annual Reports.*
18. A United News of India Record (1975), *Aspects of Agriculture in India*, pp. 111-121.
19. National Council of Applied Economic Research (1974), 'Fertilizer Use on Selected Crops in India', pp. 10-52.

3

Area Concentration and Growth Rates in Agriculture

National and Inter-State Variations

3.1 INTRODUCTION

A STUDY of the growth rates in agriculture in the past is important in many ways. First, such a study would enable an assessment of the progress of agricultural development in terms of the various components of agricultural structure and their positive and negative contribution to the growth of agriculture. Such a study would however be incomplete if it does not take into account the regional variations of agricultural structure and the processes at work to influence variations in growth rate of agriculture between regions.

Due to sharp differences in topography, soil, climate and water availability between regions, the development of agriculture is hindered. Therefore, a knowledge of the relative performance of different regions in the past will be useful in deciding priorities for the future.

This Chapter undertakes to persent average growth rates in area, output and yield of some major crops first, at the national level and than for seventeen States covering most parts of the country for the decade 1960-70. This period has been divided into three time blocks. *Time Block I* respresents three year average for 1960-61, 1961-62 and 1962-63. *Time Block II* represents four year average for 1963-64, 1964-65, 1965-66 and 1966-67. *Time Block III* represents the average for 1967-68, 1968-69 and 1969-70.

Crop area concentration—National level

At the national level, gross cropped area increased by 1.17 per cent and 4.5 per cent during 1963-66 and 1967-70 respectively, with the average of 1960-63 as base. Rice occupies the first place in all the time blocks. It is followed by Jowar, Wheat, other pulses and bajra respectively. This indicates the importance of food crops in the country. If all the food crops are taken into consideration, the percentage of gross cropped area under them works out to 78 per cent during 1960-63, 76.7 per cent in 1967-70 and 75 per cent in 1970-72. During 1970-72, the relative position of different crops remained more or less unaltered over the earlier years, except for rice and wheat. The percentage of gross cropped area under rice increased from 22.88 per cent during 1967-70 to 24.5 per cent during 1970-72. The percentage of gross cropped area under wheat increased from 9.76 per cent during 1967-70 to 11.36 per cent during 1970-72. This is partly due to the fact that wheat cultivation has spread to the hitherto rice growing regions also specially after 1966, the year, in which wheat revolution had just started in Punjab-Haryana region. During 1967-70, the percentage of gross cropped area under wheat increased to 9.76 per cent from 8.34 per cent in 1963-67, over the base 1960-63. Wheat and Rice areas are however mutually exclusive, if we take into consideration the overall spatial pattern of concentration with special reference to agro-climatic conditions favourable to these crops. Relatively speaking, rice cultivation has continued to concentrate in the rice dominant areas of humid conditions (rainfall over 80 inches). Wheat, on the other hand, is concentrated in the alluvial plains of the north and some areas of the black soil plateau where irrigation and/or winter rains are adequate.

Among the non-food crops groundnut and cotton are relatively more important in all the time blocks. During 1970-72 also, they are relatively more important in terms of percentage of gross cropped area under them. The relative position of crops however has not changed significantly between 1960-72, thus leaving the cropping pattern almost unchanged.

3.2 GROWTH RATES IN AREA, OUTPUT AND YIELD—NATIONAL LEVEL

Growth Rate in Area

During the decade 1960-70, there has been a significant increase in area under rice, wheat, bajra, maize, sugarcane and potatoes. During 1967-70, the highest growth rate in area is recorded by sugarcane followed by potatoes, maize and wheat respectively. Bajra, dry chillies and rice are some other crops whose area has increased. Between rice and wheat, the area under wheat increased more than that of rice during the same period. (18.5 per cent under wheat as against 7.5 per cent under rice). Between 1960-63 and 1970-72, area under rice increased by 10 per cent as against 44.6 per cent in the case of wheat which is partly due to changes in cropping pattern in favour of wheat in the predominantly wheat growing areas and partly due to extension of wheat cultivation in the adjoining rice growing areas such as in Bihar, West Bengal and so on. In the case of rice, however, growth rate in area is comparatively less widespread.

3.3 GROWTH RATES IN CROP OUTPUT—NATIONAL LEVEL

Wheat, potatoes, maize and bajra respectively, recorded highest increase in production during 1967-70. The production of many crops which had declined during 1963-67, recovered during 1967-70. An interesting feature to note is that production of rice declined during 1963-67 and 1967-70. Between 1960-63 and 1967-70 wheat production showed the highest rate of growth (63.4%). This was due to increase in acreage and yield of wheat during the period 1967-70. Sugarcane, potatoes and maize whose output growth was high also had sharp increases in area under them during the period 1967-70. The sharp decline in production was in the case of Mesta followed by rice and linseed. Among other crops whose production declined during the period 1967-70 are pulses, gram, small millets, ragi and jute. In the case of these crops a fall in production may partly be associated with a decline in area. However, cotton and barley production increased inspite of decline in area.

3.4 GROWTH RATES IN YIELD—NATIONAL LEVEL

Between 1960-63 and 1967-70 the yield of wheat, cotton, tobacco, sugarcane and jowar increased significantly. Wheat yield increased by 35 per cent between 1960-63 and 1967-70, whereas rice yield increased by 7 per cent. The yield decline was highest for mesta (this perhaps explains partly the highest decline in mesta output). Other crops whose yields declined were rape and mustard, potatoes and chillies. In the case of potatoes, the increase in production was highest after wheat. This could almost entirely be attributed to increase in area. The production of rice declined inspite of an increase in area and very nominal increase in yield.

In case of yields too, one finds that most of the crops with declining yields between 1960-63 and 1963-67 improved their position between 1960-63 and 1967-70.

3.5 AREA CONCENTRATION AND GROWTH RATES IN AREA OUTPUT AND YIELDS—INTER-STATE COMPARISON

Only seventeen States have been included in this analysis. Uttar Pradesh, Madhya Pradesh and Maharashtra account for highest percentage of cropped area in India. These are States with relatively larger geographical area. However, in each State, percentage of cropped area to total geographical area has remained almost unchanged.

Inter-State variations in cropped area bring out broadly the relative importance of the Indo-Gangetic plain extending from Punjab-Haryana through Uttar Pradesh and Bihar to West Bengal. Next in importance are the littoral States with extensive coastal plains as in Kerala (61%), black soil areas and deltaic landscapes of Tamil Nadu, Andhra Pradesh, Karnataka and Gujarat. Rajasthan offers a contrast in that the cultivated area is more and yet much of it comprises dry sandy areas or rocky areas of the Aravalli Hills. Assam and Orissa are typical of rugged terrains with forests occupying a larger portion of the geographical area.

3.6 AREA UNDER EACH CROP IN THE STATE TO GROSS CROPPED AREA UNDER THE STATE

Rice States

The southern region consists of Andra Pradesh, Tamil Nadu, Karnataka and Kerala. In Andhra Pradesh area under rice as a percentage of total cropped area is the highest during 1960-63 to 1967-70. Rice is followed by jowar, pulses, small millets and groundnuts in area concentration.

In Tamil Nadu, Rice, Jowar, Bajra, Cotton and Groundnuts are important in area concentration.

Karnataka has highest area concentration in the case of Jowar followed by rice, ragi, groundnuts, cotton and pulses. Kerala has the highest area concentration in rice.

In the Southern States, rice and jowar are the most important crops in area concentration. In Andhra, Tamil Nadu and Karnataka groundnut is an important commercial crop.

Eastern Region consists of Bihar, Assam, West Bengal and Orissa. In all these States rice is predominant in area concentration. In Bihar, rice is followed by pulses and maize in area concentration. In Assam, rice occupies 73.8 per cent of total cropped area. In West Bengal ricc is cultivated on 71 per cent of gross cropped area and pulses and jute are other important crops in area concentration. In Orissa, rice is the predominant crop in area concentration. It is followed by pulses.

These eight States are predominantly rice regions. The relative position of rice remained unchanged in area concentration between 1960-70.

Wheat States

In Punjab and Haryana, wheat is the premier crop occupying 37.28 per cent of total cropped area in 1967-70 and it is followed by maize, gram, cotton and rice respectively.

In Uttar Pradesh area under rice was 19.79 per cent of total cropped area in 1967-70 and area under wheat was 22.90 per cent. Eastern U.P. plain is predominantly a rice growing area while western U.P. plains cultivate wheat. Wheat and rice are followed by pulses, gram, mustard, and sugarcane.

In Madhya Pradesh, area under rice was 19.77 per cent of gross cropped area and that of wheat was 13.65 per cent in 1967-70. Pulses and gram are other important crops of the region.

In Himachal Pradesh and Jammu and Kashmir, wheat and rice are important in area. Rice was cultivated on 23.3 per cent of gross cropped area and wheat on 20.75 per cent of cropped area in 1967-70. In Himachal Pradesh ,wheat is cultivated on 39.1 per cent of cropped area and rice on 9.0 per cent of cropped area in 1967-70.

In Rajasthan, bajra accounts for 30.5 per cent of gross cropped area, and it is followed by pulses, wheat and gram.

Finally, in Western India consisting of Maharashtra and Gujarat we find that neither rice nor wheat is of primary importance in area concentration. In Maharashtra, jowar has the highest area followed by cotton, bajra, rice and pulses. In Gujarat groundnut has highest area under it followed by bajra, cotton and jowar.

Both these States are different from the other States in the sense that commercial crops are more important in terms of percentage of total cropped area under them.

Except in Gujarat and Maharashtra where area under cotton and groundnut is higher, in all other States, food crops predominate. In Maharashtra, the first place goes to Jowar. There are however States like Tamil Nadu, where during 1967-70 groundnut and cotton had gained more area under them. Similarly, jute in West Bengal ranks third after rice and other pulses. In these States, and to some extent in Karnataka and M.P., there is a trend towards changes in percentage of total cropped area devoted to some of the crops. In States where food crops have always predominated, there is hardly any change in the percentage of gross cropped area devoted to them. This is bound to be true in a country where food crops dominate the agricultural scene, and are mostly for self-consumption.

Secondly, except in Rajasthan, Maharashtra and Gujarat, in all other States, rice occupies a fairly significant place. Similarly in the case of wheat, except Tamil Nadu, Assam, Orissa, West Bengal, Karnataka and Andhra Pradesh in all other States it occupies an important place in terms of per cent of gross cropped area devoted to it. This is not surprising for, rice, jowar, wheat and bajra account for almost 50 per cent of the country's gross cropped area.

3.7 RELATIVE SHARE OF EACH STATE TO THE TOTAL AREA UNDER IMPORTANT CROPS

Most of the rice acreage is accounted for, by Bihar, West Bengal, Madhya Pradesh, Uttar Pradesh, Orissa, Andhra Pradesh and Tamil Nadu respectively. These States accounted for 77 per cent of national area under rice, in 1967-70.

In the case of jowar, Maharashtra, Karnataka, Andhra Pradesh, Madhya Pradesh and Gujarat account for most of the area under jowar in the country as a whole in all the three time blocks. They account for 76.68 per cent of national area under jowar in 1967-70. In the case of Bajra, Rajasthan accounts for 39.05 per cent, 40.05 per cent and 36.90 per cent respectively, of the area under bajra during 1960-63, 1963-67 and 1967-70. This State is followed by Maharashtra, Gujarat and Madhya Pradesh respectively. These States together accounted for 79.12 per cent of national area under bajra.

In the case of wheat, Uttar Pradesh accounts for, on an average 35 per cent of wheat area in the three time blocks, followed by Madhya Pradesh, Punjab, Haryana, Rajasthan. These States together accounted for 87.7 per cent of national area in 1967-70. Taking some of the important non-food crops into consideration, Maharashtra and Gujarat account for most of the area under cotton in India. They are followed by Karnataka and Madhya Pradesh. These States account for 79.28 per cent of national area under cotton.

In the case of groundnut, Gujarat and Maharashtra account for 50 per cent, 55 per cent and 38 per cent respectively during 1960-63, 1963-67 and 1967-70. Their share in groundnut acreage has declined followed by Karnataka, Tamil Nadu and Andhra Pradesh and they account for 38 per cent, 40 per cent and 42 per cent of the total area under groundnut in India during 1960-63, 1963-67 and 1967-70 time blocks respectively. These five States together accounted for 80 per cent of national area under groundnut. Uttar Pradesh accounts for nearly 50 per cent of the sugarcane (*Gur*) area in these time blocks.

This State is followed by Punjab and Haryana, Maharashtra, Bihar, Karnataka and Andhra Pradesh respectively. These together accounted for 88 per cent of national area under sugarcane.

In the case of tobacco, Andhra Pradesh followed by Gujarat and Karnataka account for 69.20 per cent of the national area under the crop.

3.8 GROWTH RATES IN AREA

Rice States

In Andhra, the percentage of area under rice showed an increase of 1.9 per cent per annum during 1960-63 to 1963-67, but between 1960-63 and 1967-70, there is a decline of 0.90 per cent. The other crops whose area has increased during the same period are sugarcane, dry chillies, tobacco and groundnut. These have registered significant increase both in 1963-67 as well as in 1967-70, over the 1960-63 level. Area under cotton, gram and bajra declined during this period.

Tamil Nadu

In Tamil Nadu, sugarcane is the only crop which increased in area significantly between 1960-63 and 1963-67. Rice, cotton, groundnut and dry chillies also increased, but not very significantly. In contrast during 1960-63 and 1967-70 again sugarcane increased significantly, while crops like gram, maize and potatoes also increased significantly. Position of rice improved slightly.

Karnataka

In Karnataka, rice, cotton, groundnut, maize, sesamum and mesta increased in area, during 1963-67 as compared to the 1960-63. During 1967-70 rice, bajra, gram, cotton, maize, sesamum, linseed, mesta and potatoes increased in area. In particular, maize and gram made a significant improvement between 1960-63 and 1967-70.

Kerala

The cropping pattern in Kerala is less diversified than in most other States included in this study. An important crop in the State but not included in our analysis is coconut. Rice is the only crop which registered a 3 per cent increase between 1960-63 and 1963-67 and between 1960-63 and 1967-70 it increased by 11.7 per cent. Sesamum is another crop which registered a 9 per cent increase in area between 1960-63 and 1967-70.

Bihar

During 1963-67 jowar, bajra, groundnut, wheat, linseed and potatoes are the crops that registered increases. The rest of them declined. The area under jowar, bajra and groundnut increased significantly, and the area under wheat increased nominally. During 1967-70 jowar, wheat, bajra, groundnut, maize, tur, sesamum and potatoes increased in terms of the per cent of area under them. Significant among them were jowar, groundnut, wheat, sesamum and potatoes. In Bihar one finds that in 1963-67 and 1967-70 once again the percentage of area under most crops increased.

Assam

In this State, during 1963-67 wheat, pulses, cotton, maize, small millets, mustard, sesamum, sugarcane, jute, mesta and potatoes increased. The only crops to have registered a decline in area were rice and gram.

During 1967-70, jute was the only crop which registered a decline in area. The rest of the crops registered an increase. Significant among these were rice, wheat, maize, tur and potatoes. Thus, in Assam most of the crops grown in the State have increased in area over the first time block (1960-63).

West Bengal

During 1963-67 rice, wheat, other pulses, maize, small millets, rape and mustard, tur, sesamum, jute, chillies and potatoes

toes increased in area. Increase in area under maize, small millets, sesamum, tur, jute and chillies was particularly significant. During 1967-70 wheat, rice barley, small millets, mustard, sesamum, jute, chillies and potatoes increased in area. Increase in area was significant during this period in the case of wheat, small millets, sesamum, jute and barley.

Orissa

During 1963-67 all crops except jowar, bajra, gram, maize increased in their area over the first time block (1960-63). Most crops increased in area significantly. In this respect Orissa is exceptional. During 1967-70 bajra, declined very sharply in area and all the other crops increased very significantly.

Wheat States

Punjab and Haryana

During 1963-67 rice, wheat, jowar, gram, cotton, groundnut, barley, tur, sugarcane and potatoes increased in area. But significant increase in area was noticed in the case of groundnut and tur, followed by potatoes.

During 1967-70 however, rice, wheat, groundnut, barley and tur increased in area. Very significant increases were registered in the case of tur, groundnut, barley, wheat and rice.

Uttar Pradesh

The percentage of area under jute, wheat, pulses, groundnut, small millets, dry chillies, linseed and potatoes increased significantly in Uttar Pradesh during 1960-63 to 1967-70. For most other crops, area declined.

Madhya Pradesh

During 1963-67, percentage of area under all crops except, wheat, gram, barley, mustard, sesamum, linseed, castor and tobacco increased with the average of 1960-63 as base. In the case of bajra, tur and mesta, increase in area was more significant. During 1967-70, however, percentage increase in area was registered in the case of important food crops like rice, jowar and bajra. Between 1963-67 and 1967-70 mesta and bajra have shown significant increases in area over the first block (1960-63).

Himachal Pradesh

In Himachal Pradesh, rice, maize, small millets, barley, rape and mustard, sugarcane and potatoes increased in area during 1963-67 over 1960-63 level. Significant of these were rice, other pulses, maize, mustard and sugarcane.

During 1967-70 with the average of 1960-63 as base, rice, wheat, other pulses, maize, barley, small millets, mustard, sesamum, linseed, sugarcane and potatoes increased in area. Significant among these were, rice, wheat, other pulses, maize, barley, rape and mustard and sugarcane.[1]

Jammu and Kashmir

In Jammu and Kashmir during 1963-67 rice area increased negligibly. Other crops that showed increase were wheat, other puises, maize, rape and mustard and sugarcane. Significant among these were other pulses, rape and mustard and sugarcane. During 1967-70 percentage of area under rice, wheat, rape and mustard and sugarcane increased over 1960-63 level. Of these, percentage increase in area under rice, rape and mustard and sugarcane was most significant.

Rajasthan

In crops which increased in area were groundnut, cotton, sesamum, tobacco and tur during 1963-67 over 1960-63 level.

During 1967-70 groundnut, rice, jowar, bajra, pulses, cotton, maize, barley, small millets, sesamum, chillies and potatoes increased in area but of these significant were percentage increase in area under groundnut, potatoes, maize and cotton.

Gujarat

During 1963-67 jowar, wheat, bajra, cotton, groundnut, maize, sesamum, sugarcane and potatoes increased in area. Significant among these were, wheat, sesamum, sugarcane and potatoes. During 1967-70 maize, ragi, tur, sesamum, sugarcane, chillies and potatoes increased in area. Significant of these were wheat, bajra, maize, sugarcane, potatoes and to some extent ragi and sesamum. The area under cotton declined between 1960-63 and 1967-70.

Maharashtra

In this State, maize, cotton and bajra registered maximum increases in area between 1960-63 and 1967-70. Rice area also increased. There was a decline in area under sugarcane, groundnut.

The analysis above, of percentage variations in area under different crops during 1963-67 and 1967-70 shows slight changes in cropping pattern over 1960-63 level. There are States like Orissa where area under all the crops has increased. This can be attributed to increase in total cropped area and extension of irrigation.

Between the time blocks 1963-67 and 1967-70 variations are not so marked. A point worth noting is that in almost all the States food crops have also increased in area though not significantly throughout. This shows that area devoted to food crops has not changed significantly enough to show a marked change in our traditional cropping pattern.

In the case of wheat and rice in most States crops are grown for historical reasons and a large part of this is meant for self-consumption, therefore, their relative position over the three time blocks (1960-63, 1963-67 and 1967-70) has remained largely unchanged. In Himachal Pradesh and Jammu and Kashmir area under this crop has increased significantly in 1967-70 with 1960-63 as base. These two States are followed by Punjab and Haryana. Growth rate in rice area is significantly higher in 1967-70. In Punjab and Haryana, rice yields have also shown a similar trend as will be seen in subsequent analysis.

Similarly, in the case of wheat, Assam, Orissa and West Bengal have shown significant increases in area relatively. Gujarat has also shown a significant percentage increase in area under wheat.

Among the non-food crops, cotton, groundnut, sugarcane, mesta, sesamum and rape and mustard have increased in area in most States.

3.9 GROWTH RATES IN OUTPUT AND YIELD PER HECTARE

Rice States

Southern Region

Andhra : In 1967-70 with 1960-63 as base, production and yield of rice, tobacco, groundnut, maize, tur and chillies have increased in Andhra. The maximum increase in production was recorded by groundnut (59%), followed by sugarcane (40.5%), tur (39%), maize (38%) and tobacco (30%). The yield per hectare of groundnut increased by 10 per cent only, while the area under groundnut increased by 58 per cent. The same is true in case of tur, sugarcane and tobacco. In the case of rice, yield per hectare increased by 11 per cent and production by 28 per cent, whereas area under it had fallen by 0.80 per cent in 1967-70.

Tamil Nadu : In Tamil Nadu again, yield per hectare has remained low or negative for most crops. The only exception being rice (+7%), potatoes (+11%), cotton (+9%), bajra (+2.5%) and sugarcane (+1%). The increase in production is significant for gram, barley, sugarcane and potatoes and lastly for rice.

Karnataka : In Karnataka, also, maize showed highest rate of growth in production accompanied by significant increase in area and yield. The production of many crops declined. Growth rate in yield was significant in the case of maize, gram, wheat, rice, tur and castor.

Kerala : In Kerala, production of rice, groundnut and sugarcane increased significantly compared to the other crops. Rice production increase was associated with increase in area and yield. In the case of groundnut, yield increased by 99 per cent whereas area declined by 19 per cent. The production of cotton and tobacco declined due to fall in area and yield.

Eastern Region

Bihar : The State recorded highest increase in production in the case of mustard (22%) during the period 1967-70 with 1960-63 as base, followed by wheat, cotton, sesamum and potatoes.

Assam : In Assam, the production of linseed in particular increased significantly in 1967-70 and was followed by maize, wheat and small millets. Rice production fell by 9.6 per cent inspite of an increase in area by 15.6 per cent and 6 per cent increase in yield. In general, yield increases remained insignificant in the State.

West Bengal : The production of sesamum, tur, ragi, small millets, barley, maize, gram, other pulses, jute, chillies and wheat increased. The highest growth rate in yield was for potatoes (67%), followed by tur (28%), linseed (27%), maize (17.3%) and rice (5.7%). Wheat yield declined by 1 per cent during 1963-67. But during 1967-70 the production and yield of wheat increased very significantly. In this time block, rice, other pulses, gram, maize, barley, registered increases in both output and yield. In the case of sesamum, ragi, small millets, and jute increase in production is explained more by increase in area as the yield has remained rather low. In the case of wheat, high rate of growth in output and yield is explained by the fact that the crop is of relatively recent origin in the State.

Orissa : In Orissa, there has been significant growth in production in almost all the crops. Cotton and linseed are not widely grown because of unfavourable conditions and hence the decline in production is to be expected. Similar is the case with decline in yield per hectare in mesta, rape and mustrad and small millets.

Wheat States

The maximum increase in production was recorded in Punjab and Haryana in the case of tur. Other crops whose production increased significantly were rice, wheat, bajra, groundnut, maize, barley and sugarcane. Yields have remained low in case of tobacco, mesta, linseed, sesamum, tur, ragi, small millets and jowar. In the case of tur, area has increased significantly whereas yield has remained low. The cropping pattern productivity and their changes over the years do reveal that this region is ahead of others leading to specialisation of crops due to which number of crops grown tends to be less.

Madhya Pradesh

The production of rape and mstard and ragi by 112.5 per cent and 100 per cent respectively. In both cases, yield declined and area increased by 3.2 per cent and 60 per cent respectively.

Madhya Pradesh like Assam is characterised by low or declining growth rate in yield of most crops during the period 1967-70.

Himachal Pradesh

Production of ragi, linseed, gram, barley, maize, potatoes, rice and wheat increased significantly. In the case of both wheat and rice area increase was more significant than yield increase.

Jammu and Kashmir

Wheat, rape and mustard, other pulses, sugarcane and rice showed highest rates of growth in production. The yield per hectare of these crops increased and were associated with increases in area too. Rice area increased by 83.2 per cent and yield by 29 per cent.

Rajasthan

The production of maize, castor, tobacco, barley, groundnut, cotton and wheat increased. Area increases were more significant. Yield growth rates have been low or negative for most crops.

Gujarat

In the State, Potatoes were followed by wheat, bajra,

sesamum and sugarcane, in recording highest increase in production during the period 1967-70. The production of rice and many other crops declined. In the case of potatoes, wheat and bajra, yield also increased. In the case of rice both area and yield declined.

Maharashtra

Rice production increased rather significantly in 1967-70 (40.3%) over the base year 1960-63. This was accompanied by only 4.3 per cent increase in area and 2 per cent increase in yield. Bajra, sugarcane and sesamum were other crops to register significant increases in production. Slight decline in area and yield of sugarcane is difficult to explain with aggregate data particularly when we find that Maharashtra is leading in sugarcane cultivation and the yields are the highest in the country.

The highest increase in production and yield was recorded by maize in Maharashtra. Except maize, and other crops mentioned above, yield increase was rather insignificant.

During 1967-70, in most States, the increase in output is significant. Growth rates in yield are however not very significantly different especially in the case of major food crops like rice and wheat in the traditional rice and wheat cultivating states. But in states where the two crops are grown on a small percentage of gross cropped area the increase in yield was significant, e.g., wheat yields in West Bengal and rice yields in Punjab. These were the innovations taking advantage of irrigation and remunerative price for superior variety of rice in Punjab and improved variety of wheat suited to local condition in West Bengal.

But overall trend in yield is one of decline particularly during 1963-67 with the average of 1960-63 as base.

3.10 NATIONAL AND INTER-STATE VARIATIONS IN AGRICULTURE—AN OVERVIEW

The percentage of area devoted to various crops has hardly changed between 1960-70 at the national level. Crop area concentration has however changed in some States like Maharashtra and Gujarat where two important commercial crops cotton and groundnut are gaining importance. Tamil Nadu, Karnataka and Andhra Pradesh are three other States where groundnut is an important crop. These five states together account for nearly 80 per cent of national area.

However, food crops predominate in all the states except Gujarat where area under groundnut and cotton is higher than under bajra, in terms of percentage of gross cropped area devoted to it. In southern and eastern parts of India including Orissa and Eastern Uttar Pradesh rice cultivation predominates even now and in the northern states, wheat, gram and pulses predominate. In Central India, rice, wheat and pulses are the predominant crops in terms of area concentration. Crops like tobacco and chillies which are important from commercial point of view, in Andhra Pradesh, have a small percentage of total cropped area under them.

By and large, the crop pattern in India in terms of proportion of area devoted to crops is traditional with food crops accounting for 75 per cent of total cropped area at the national level of which cereals account for 61.2 per cent and pulses 13.8 per cent in 1970-72, and rice and wheat occupy the first two places.

In terms of percentage growth rate in area of different crops between 1960-63 and 1967-70, there are marked inter-state variations. In some States like Orissa, area under all food crops has increased which is associated with increase in area brought under cultivation and provision of extended irrigation. In almost all the states area under food crops increased during this period. But between 1963-67 and 1967-70, variations are not marked. It is interesting to observe that rice area has increased in Punjab and Haryana, Himachal Pradesh and Jammu and Kashmir while wheat area has increased in Assam, Orissa, West Bengal and Gujarat between 1960-63 and 1967-70. Wheat cultivation in now penetrating into the hitherto rice growing regions. Among the non-food crops, growth rates in area of cotton, groundnut, sugarcane, mesta, sesamum, rape and mustard are quite significant in most states. Finally, between 1960-63 and 1967-70, in most States increase in output is significant while yield growth rates are not very significantly different specially, for food crops (Rice and wheat) in the traditional rice and wheat cultivating regions. However, yield variations are more marked in the areas where rice or wheat are of more recent origin. For example, in West Bengal, wheat yields have increased significantly. In Punjab and Haryana rice yields have increased very significantly.

In Punjab and Haryana, yields have increased between 1960-63 and 1967-70 for all the crops. In this respect, this region is exceptional. The yield growth rates are positive for most other crops. This confirms the well established belief that the impact of green revolution was felt first in Punjab-Haryana region as early as 1965-66.

REFERENCE

1. For linseed and sesamum the base selected was (1963-66) second time block, as data for the first time block (1960-63) were not complete.

4

Spatial Concentration and Growth Rate of Rice and Wheat

An Inter-Regional Analysis

4.1 INTRODUCTION

INTER-STATE variations in area, output and yield of 23 principal crops analysed in the previous chapter serves to highlight main regional variations in agriculture. Agriculture is a spatial subject and the space element involves a number of factors—physical, economic and institutional which by their inter-relationship, contribute to the emergence of relatively homogeneous crop regions of which, rice and wheat are important constituents. Regional dimensions can be better understood and evaluated in terms of smaller administrative units than the States which are of varied size and often too heterogeneous within the data constraints and possibilities of providing regional dimension to agriculture at different scales of areas. The district is considered as an ideal unit for data computation, analysis and regional grouping based on the concept of homogeneity in one or more factors related to agriculture. This chapter attempts to present a spatial and temporal pattern of distribution of rice and wheat in India using district-wise data for four time blocks, viz., 1960-63, 1963-66, 1966-68 and 1970-72. Core and peripheral areas of rice and wheat are marked off on a map as the initial step in demarcating regions most favourable to these crops. The boundaries of such regions are then adjusted to the nearest boundary of the districts to facilitate the analysis of spatial concentration and spatial and temporal variations of rice and wheat. The importance of rice and wheat in each district was brought out by working out the proportion of area under rice and wheat to gross cropped area of the district. The relative importance of districts was brought out by ranking all of them for the four time blocks. Figures 1-4 and 8-11 bring out spatial concentration of rice and wheat for different time blocks during 1960-72. In all, 291 districts are considered for the study of spatial patterns.

4.2 SPATIAL CONCENTRATION OF RICE

Figure 1 brings out that average area concentration of rice during 1960-63 was high (above national average of 22 per cent), in parts of West Coast, East Coast, North Eastern Peninsular Plateau and the Lower Ganges Plain. During 1963-66 and 1966-68, it will be noticed from Figures 2 and 3 that area concentration remained almost the same.

During 1970-72, the area under rice to total cropped area increased to 25 per cent with infilling in the same area with the exception of Punjab-Haryana area where there has been a substantial increase in rice area.

Figure 4 shows high concentration of rice in areas which have 100 cm of rainfall. However, there is also a concentration by irrigation.

Spatial patterns and variations in yield

Rice

Figures 5, 6 and 7 bring out spatial patterns and variations in yield. Figure 5 brings out variations in average yield per hectare in 1970-72 with the average of 1963-66 as base. One hundred and twenty-six districts rank above the national averages yield of rice. Figure 6 represents the same for 1970-72 with the average of 1960-63 as base. In this period only one hundred and nine districts rank above the national average. This is due to the adoption of high yielding variety only after 1966 which brought about improvements in yield. Therefore, during

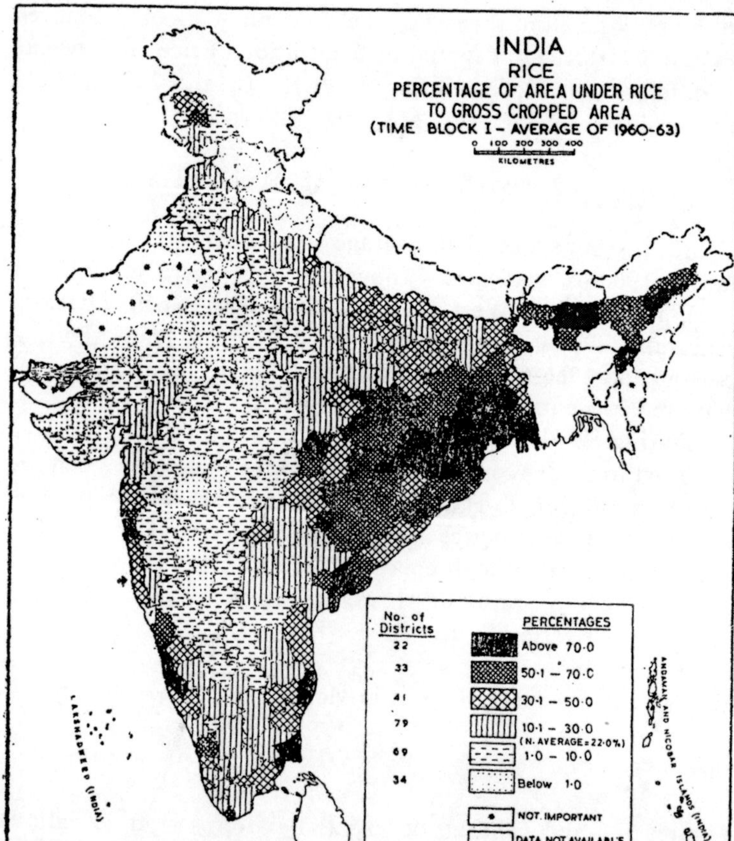
INDIA
RICE
PERCENTAGE OF AREA UNDER RICE
TO GROSS CROPPED AREA
(TIME BLOCK I – AVERAGE OF 1960-63)
0 100 200 300 400
KILOMETRES
No. of Districts
PERCENTAGES
22 Above 70·0
33 50·1 – 70·0
41 30·1 – 50·0
79 10·1 – 30·0
(N. AVERAGE = 22·0%)
69 1·0 – 10·0
34 Below 1·0
NOT IMPORTANT
DATA NOT AVAILABLE
LAKSHADWEEP (INDIA)
ANDAMAN AND NICOBAR ISLANDS (INDIA)

FIG. 1

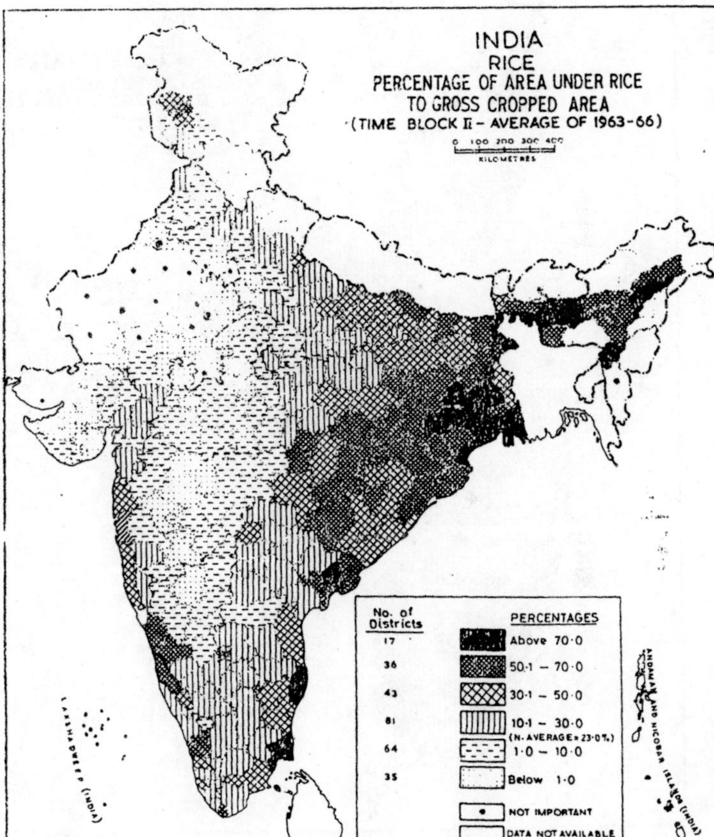
INDIA
RICE
PERCENTAGE OF AREA UNDER RICE
TO GROSS CROPPED AREA
(TIME BLOCK II - AVERAGE OF 1963-66)
KILOMETRES
No. of Districts
PERCENTAGES
17 Above 70·0
36 50·1 – 70·0
43 30·1 – 50·0
81 10·1 – 30·0
(N. AVERAGE = 23·0%)
64 1·0 – 10·0
35 Below 1·0
NOT IMPORTANT
DATA NOT AVAILABLE
LAKSHADWEEP (INDIA)
ANDAMAN AND NICOBAR ISLANDS (INDIA)

FIG. 2

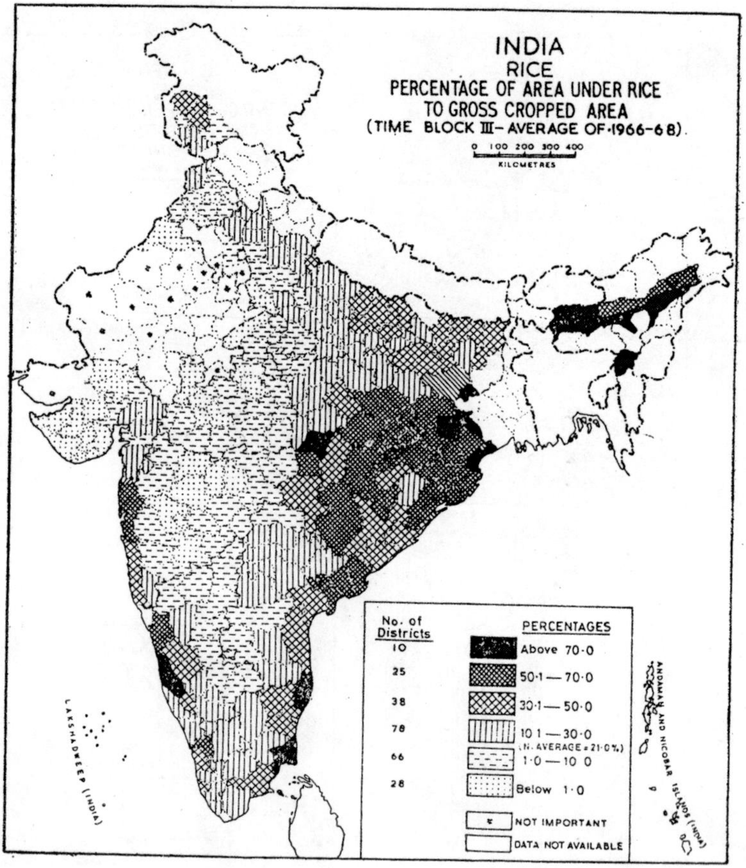
INDIA
RICE
PERCENTAGE OF AREA UNDER RICE
TO GROSS CROPPED AREA
(TIME BLOCK III- AVERAGE OF 1966-68)
0 100 200 300 400
KILOMETRES
No. of Districts
PERCENTAGES
10
Above 70·0
25
50·1 — 70·0
38
30·1 — 50·0
78
10·1 — 30·0
(N. AVERAGE = 21·0%)
66
1·0 — 10·0
28
Below 1·0
NOT IMPORTANT
DATA NOT AVAILABLE
LAKSHADWEEP (INDIA)
ANDAMAN AND NICOBAR ISLANDS (INDIA)

FIG. 3

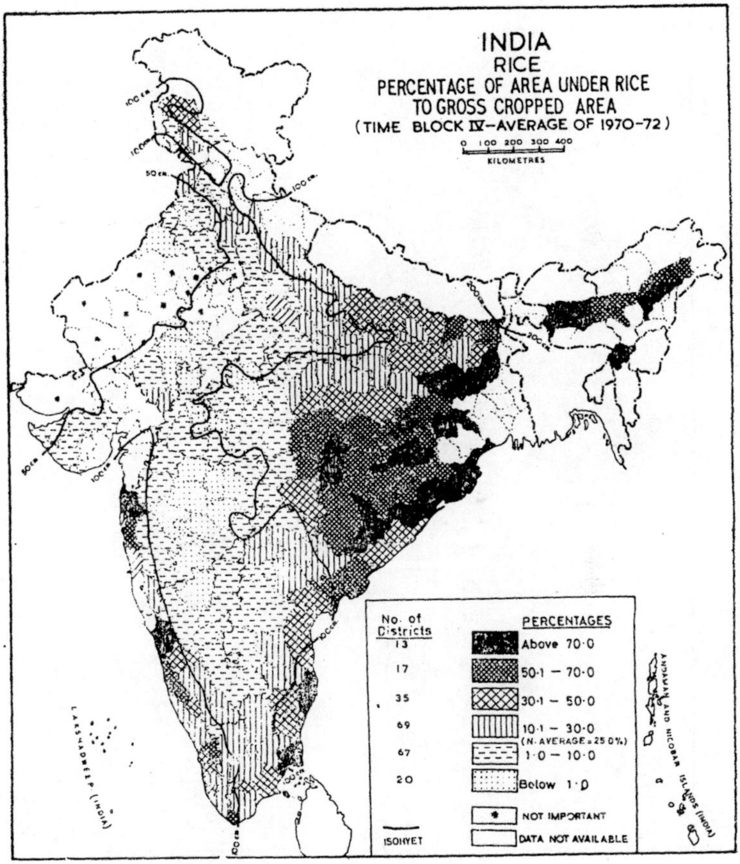
INDIA
RICE
PERCENTAGE OF AREA UNDER RICE
TO GROSS CROPPED AREA
(TIME BLOCK IV—AVERAGE OF 1970-72)
0 100 200 300 400
KILOMETRES
No. of Districts
PERCENTAGES
13 Above 70·0
17 50·1 — 70·0
35 30·1 — 50·0
69 10·1 — 30·0
(N. AVERAGE = 25·0%)
67 1·0 — 10·0
20 Below 1·0
NOT IMPORTANT
DATA NOT AVAILABLE
ISOHYET
LAKSHADWEEP (INDIA)
ANDAMAN AND NICOBAR ISLANDS (INDIA)

Fig. 4

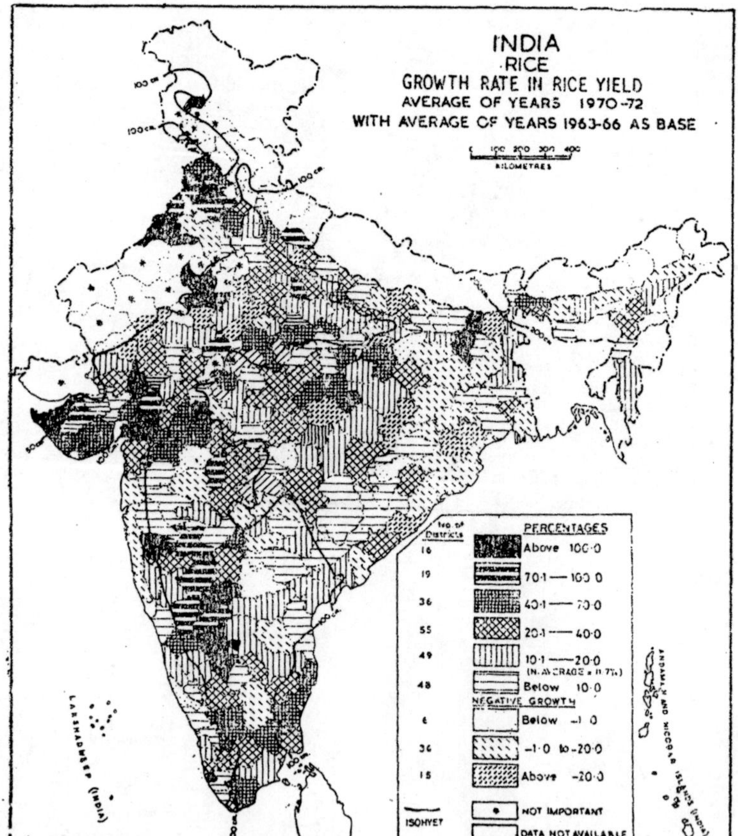
INDIA
RICE
GROWTH RATE IN RICE YIELD
AVERAGE OF YEARS 1970-72
WITH AVERAGE OF YEARS 1963-66 AS BASE
KILOMETRES
No. of Districts
PERCENTAGES
16 Above 100·0
19 70·1 — 100·0
36 40·1 — 70·0
55 20·1 — 40·0
49 10·1 — 20·0
(N. AVERAGE = 11·7%)
48 Below 10·0
NEGATIVE GROWTH
6 Below -1·0
36 -1·0 to -20·0
15 Above -20·0
ISOHYET
NOT IMPORTANT
DATA NOT AVAILABLE
LAKSHADWEEP (INDIA)
ANDAMAN AND NICOBAR ISLANDS (INDIA)
100 cm.

FIG. 5

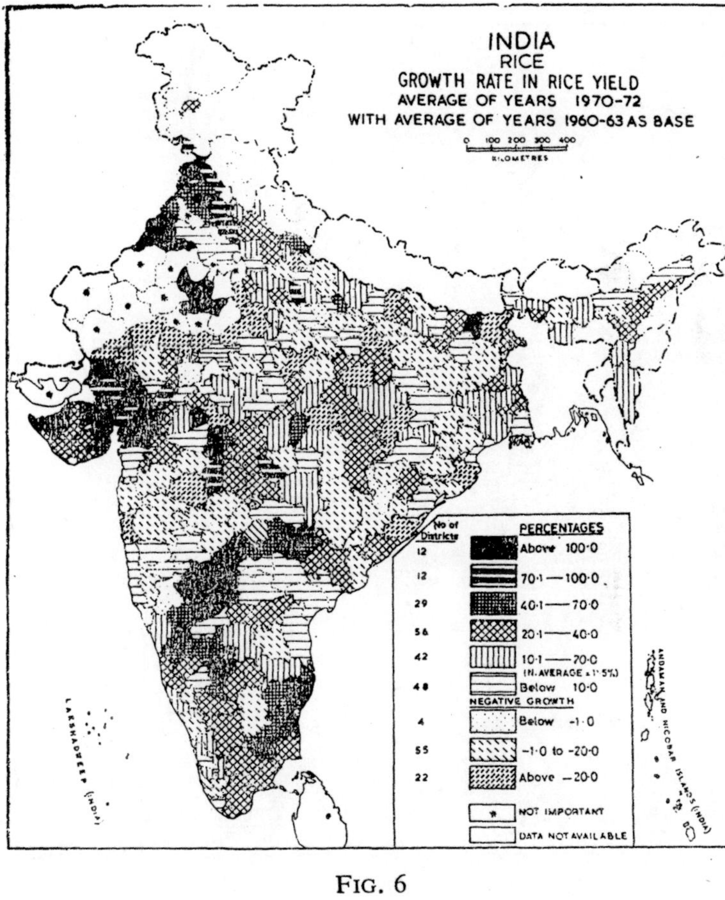
INDIA
RICE
GROWTH RATE IN RICE YIELD
AVERAGE OF YEARS 1970-72
WITH AVERAGE OF YEARS 1960-63 AS BASE
0 100 200 300 400
KILOMETRES
No of Districts
PERCENTAGES
12 Above 100·0
12 70·1 — 100·0
29 40·1 — 70·0
56 20·1 — 40·0
42 10·1 — 70·0
(N. AVERAGE = 1·5%)
48 Below 10·0
NEGATIVE GROWTH
4 Below -1·0
55 -1·0 to -20·0
22 Above -20·0
NOT IMPORTANT
DATA NOT AVAILABLE
LAKSHADWEEP (INDIA)
ANDAMAN AND NICOBAR ISLANDS (INDIA)

FIG. 6

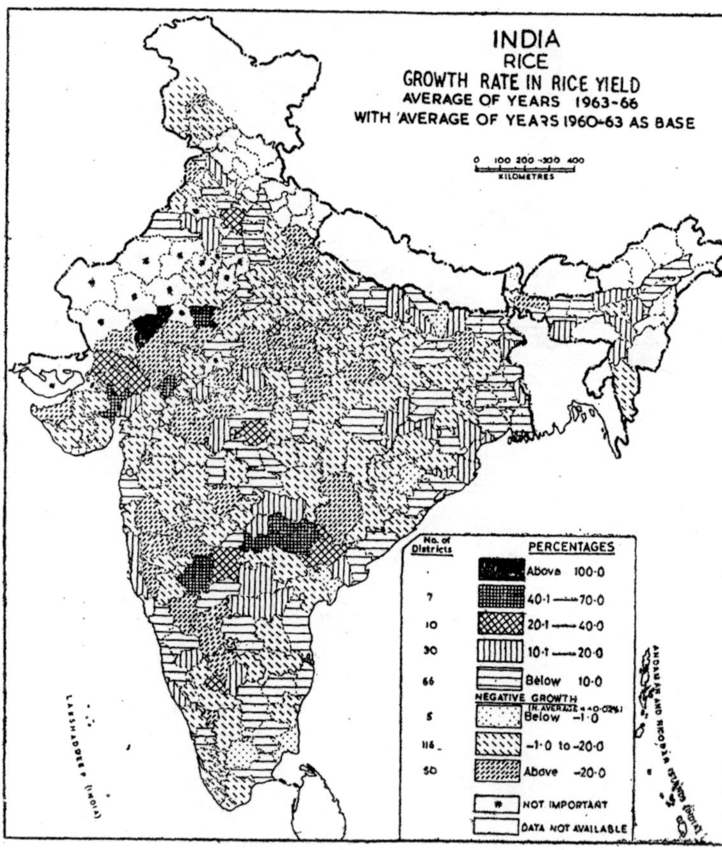
INDIA
RICE
GROWTH RATE IN RICE YIELD
AVERAGE OF YEARS 1963-66
WITH AVERAGE OF YEARS 1960-63 AS BASE
0 100 200 300 400
KILOMETRES
No. of Districts
PERCENTAGES
Above 100·0
7
40·1 — 70·0
10
20·1 — 40·0
30
10·1 — 20·0
66
Below 10·0
NEGATIVE GROWTH
5
Below −1·0
116
−1·0 to −20·0
50
Above −20·0
NOT IMPORTANT
DATA NOT AVAILABLE
ANDAMAN AND NICOBAR ISLANDS (INDIA)
LAKSHADEEP (INDIA)

FIG. 7

1970-72, with the average of 1963-66 as base, we find a change in the spread of higher yield (above national average).

Figure 7 represents yield growth rates between 1960-63 to 1963-66. In this period, the national average itself was negative (—0.2 per cent). The number of districts with negative yield growth rates was 171. In other words, low yields were a feature of most of the districts in India during 1960-63 to 1963-66.

There is a marked contrast between Figure 5 and Figure 7. In Figure 7, there are very few regions with high growth rates in yield. In 1970-72 there is an increase in the number of districts with high yields. During 1963-66 only 18 districts had a growth rate between 20-100 per cent, where as in 1970-72, 126 districts had a growth rate between 20-100 per cent. (Fig. 5). This shows a marked improvement in rice yields all over the country after the adoption of HYV.

4.3 SPATIAL CONCENTRATION OF WHEAT

Figures 8, 9, 10 and 11 bring out spatial concentration of wheat. In the case of wheat one finds that during 1960-63, higher area concentration (above national average) prevailed in the plains of Western Uttar Pradesh, Siwalik foot-hills, Central irrigated area in Punjab and Haryana, parts of Madhya Pradesh (Peninsular Plateau) and Rajasthan (Fig. 8).

During 1970-72, national average rose to 15 per cent and therefore fewer districts emerged above the national average level (Figure 11). The national average during 1960-63 was 7.2 per cent. This shows an increase of 7.8 per cent during 1970-72 over the average of 1960-63. One hundred six ranked above the national average in 1960-63 and seventy six districts were ranked above the national average in 1970-72. The number of districts in the interval between 1.00-10.00 per cent of area concentration is substantially lower in 1970-72 as compared to 1960-63, 1963-66 and 1966-68. Some of the districts had improved their position during 1970-72 as they had moved from below 1 per cent category to 1.00-10.00 per cent category.

Concentration of wheat is high in few parts where there is 100 cm of rainfall, e.g., in parts of Madhya Pradesh. Most areas cultivating wheat, however, lie in the regions of 50 cm

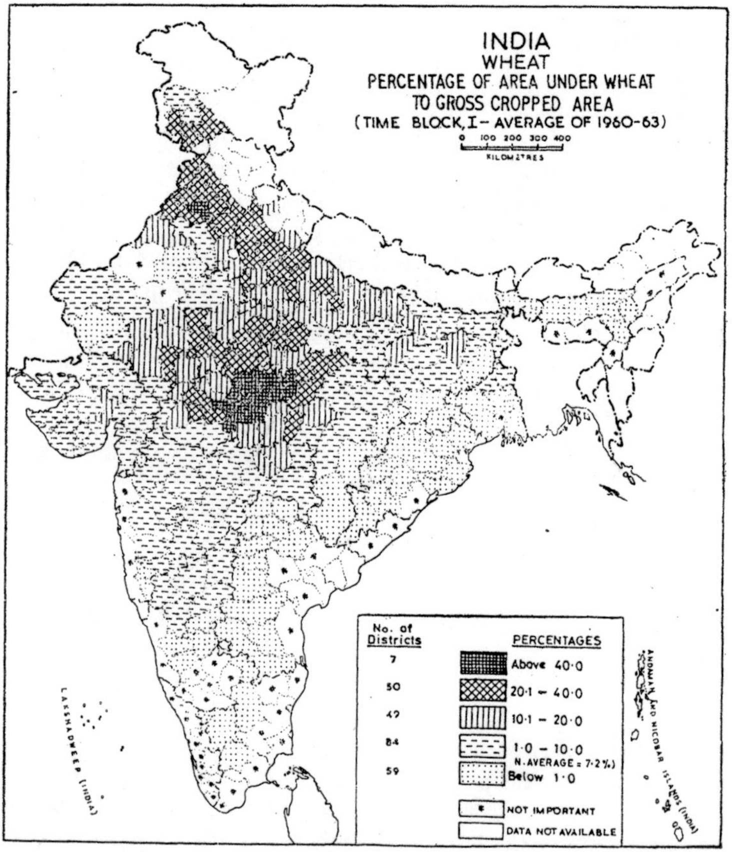
INDIA
WHEAT
PERCENTAGE OF AREA UNDER WHEAT
TO GROSS CROPPED AREA
(TIME BLOCK, I – AVERAGE OF 1960-63)
0 100 200 300 400
No. of Districts
PERCENTAGES
7
Above 40·0
50
20·1 – 40·0
42
10·1 – 20·0
84
1·0 – 10·0
N. AVERAGE = 7·2%)
59
Below 1·0
NOT IMPORTANT
DATA NOT AVAILABLE
LAKSHADWEEP (INDIA)
ANDAMAN AND NICOBAR ISLANDS (INDIA)

FIG. 8

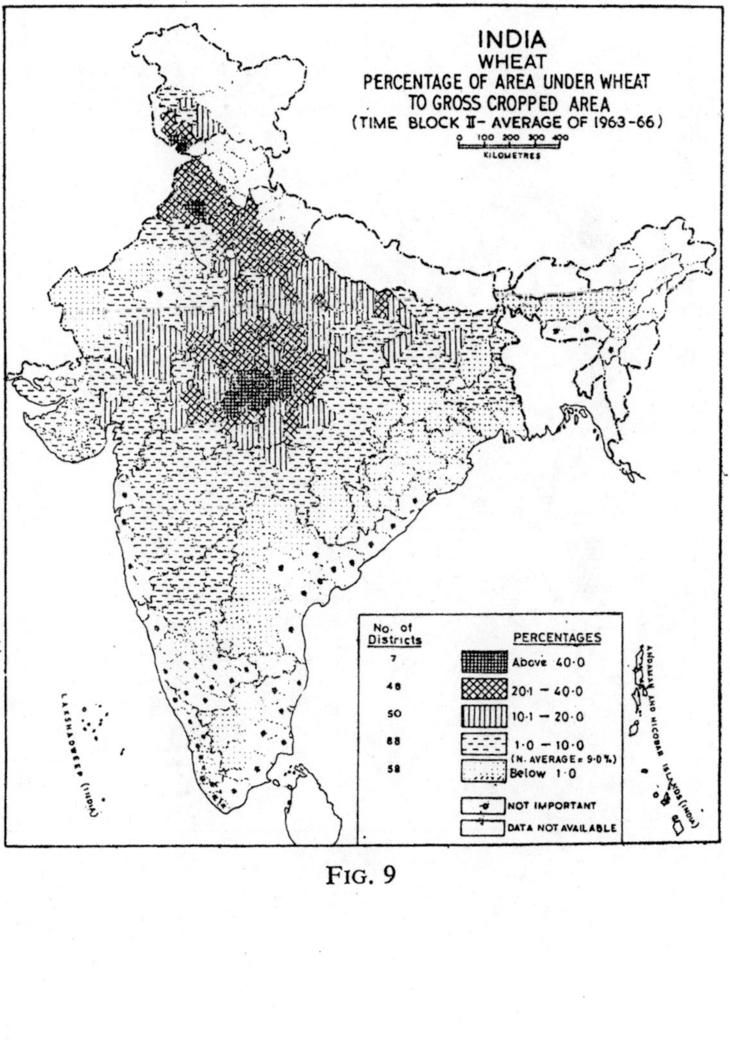
INDIA
WHEAT
PERCENTAGE OF AREA UNDER WHEAT
TO GROSS CROPPED AREA
(TIME BLOCK II- AVERAGE OF 1963-66)
0 100 200 300 400
KILOMETRES
No. of Districts
PERCENTAGES
7
Above 40·0
48
20·1 — 40·0
50
10·1 — 20·0
88
1·0 — 10·0
(N. AVERAGE = 9·0%)
58
Below 1·0
NOT IMPORTANT
DATA NOT AVAILABLE
LAKSHADWEEP (INDIA)
ANDAMAN AND NICOBAR ISLANDS (INDIA)

FIG. 9

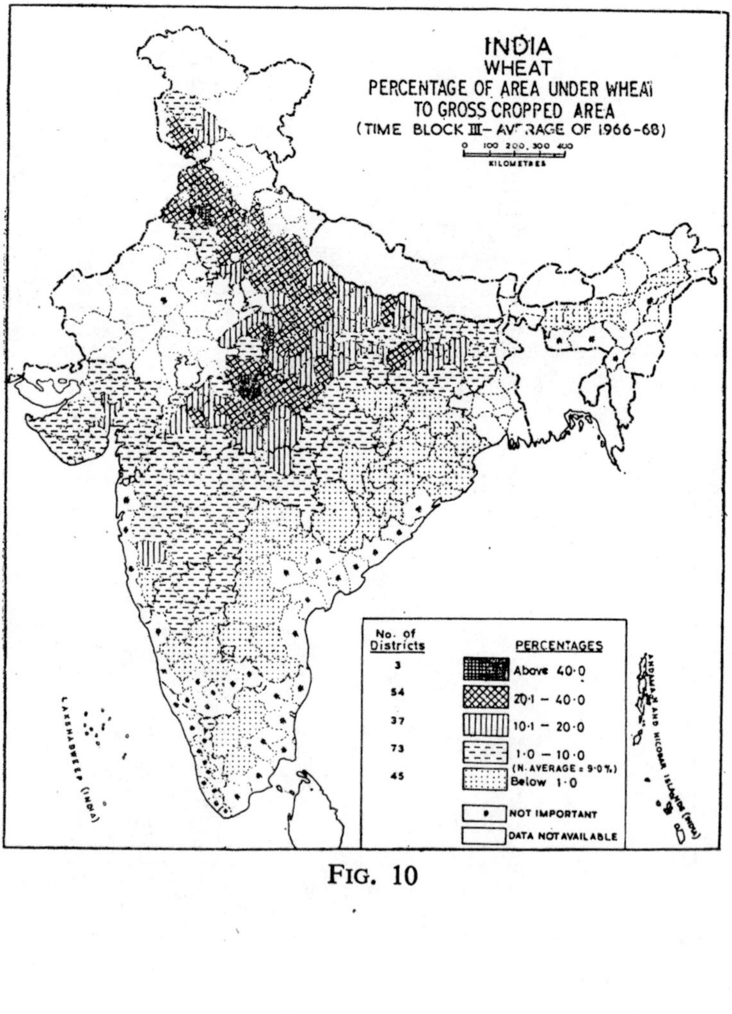
INDIA
WHEAT
PERCENTAGE OF AREA UNDER WHEAT
TO GROSS CROPPED AREA
(TIME BLOCK III– AVERAGE OF 1966-68)
0 100 200 300 400
KILOMETRES
No. of Districts
PERCENTAGES
3
Above 40·0
54
20·1 – 40·0
37
10·1 – 20·0
73
1·0 – 10·0
(N. AVERAGE = 9·0%)
45
Below 1·0
NOT IMPORTANT
DATA NOT AVAILABLE
LAKSHADWEEP (INDIA)
ANDAMAN AND NICOBAR ISLANDS (INDIA)

FIG. 10

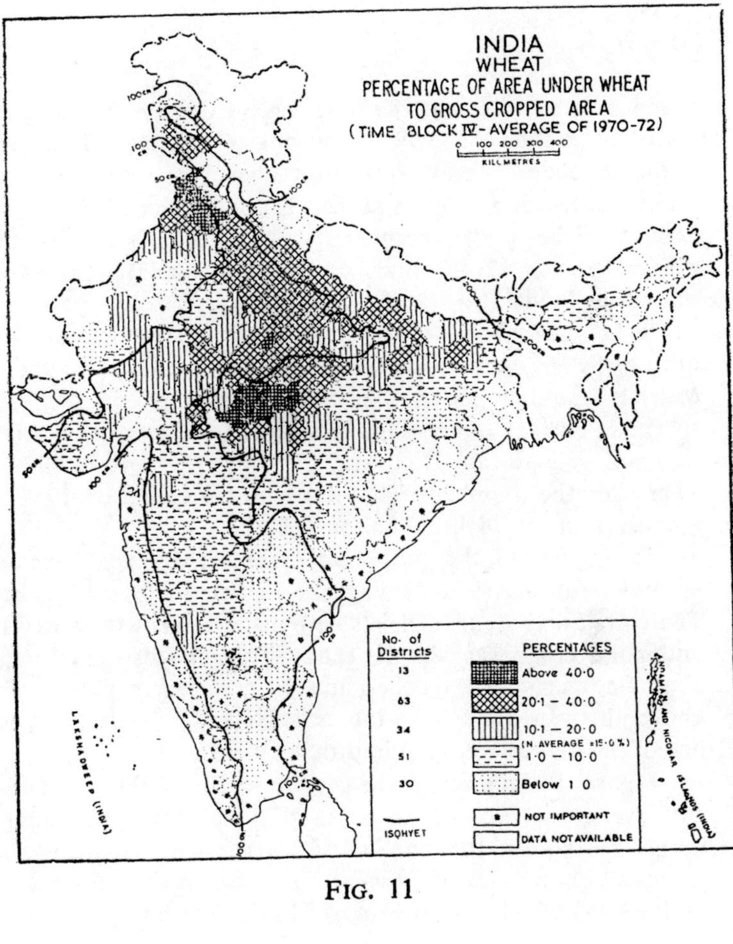
INDIA
WHEAT
PERCENTAGE OF AREA UNDER WHEAT
TO GROSS CROPPED AREA
(TIME BLOCK IV - AVERAGE OF 1970-72)
0 100 200 300 400
KILOMETRES
No. of Districts
PERCENTAGES
13 Above 40·0
63 20·1 – 40·0
34 10·1 – 20·0
(N. AVERAGE =15·0 %)
51 1·0 – 10·0
30 Below 1 0
NOT IMPORTANT
DATA NOT AVAILABLE
ISOHYET
LAKSHADWEEP (INDIA)
ANDAMAN AND NICOBAR ISLANDS (INDIA)

FIG. 11

of rainfall. In the central irrigated area of Punjab and Haryana, most of the districts have high area concentration.

Spatial patterns and variations in yield

Wheat

Growth rate in wheat yield has been extensive covering most parts of traditional wheat growing area as well as areas brought under wheat cultivation anew. In fact, spread of wheat cultivation itself was due to the impact of HYV which brought higher returns compared to other crops grown in these areas. Figures 12, 13 and 14 bring out spatial patterns and variations in yield.

Figure 12 shows that growth rates in the yield of wheat during 1970-72 with 1963-66 average as base. The number of districts ranking above the national average (62.9%) was seventy. This is interesting because in the case of rice yields, national growth rate itself was only 11.7 per cent in 1970-72. Therefore, the number of districts with above national average growth rates was higher than for wheat.

In Figure 13, the growth rates in wheat yields have been presented during 1970-72 with the average of 1960-63 as base. There are thirty four districts with negative growth rates. The national average was 58.9 per cent. Fifty eight districts (Punjab, Haryana, parts of Rajasthan and Western U.P. region) ranked above this. In Rajasthan, the extension of irrigation has been one of the major factors in improving yields.

Figure 14, represents the growth rate in yield in 1963-66 with average of 1960-63 as base. National average is negative in this period and the number of districts with negative values was one hundred seventy two. The pattern is one of low yields during 1963-66 with the average of 1960-63 as base.

4.4 THE OVERALL PATTERNS OF RICE AND WHEAT IN THE AGRICULTURAL LANDSCAPE—ANALYSIS ACCORDING TO MAJOR PHYSICAL REGIONS AND SUB-REGIONS

Having analysed the pattern of area concentration and growth in yield of rice and wheat with districts as the units, regional

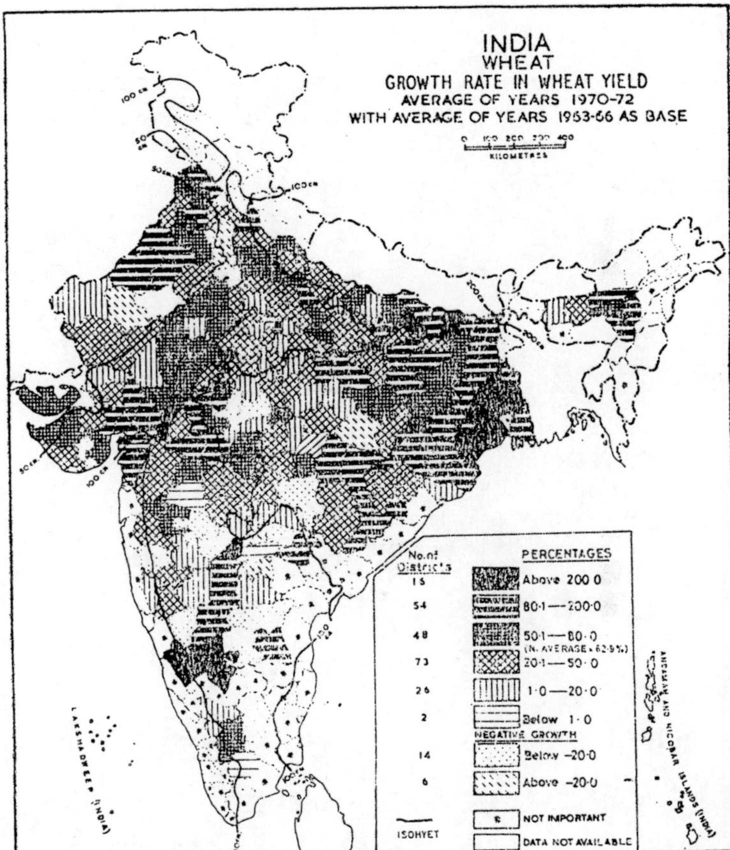
INDIA
WHEAT
GROWTH RATE IN WHEAT YIELD
AVERAGE OF YEARS 1970-72
WITH AVERAGE OF YEARS 1963-66 AS BASE
KILOMETRES
No. of Districts
PERCENTAGES
15 Above 200 0
54 80·1 — 200·0
48 50·1 — 80·0
(N. AVERAGE = 62·9%)
73 20·1 — 50·0
26 1·0 — 20·0
2 Below 1·0
NEGATIVE GROWTH
14 Below –20·0
6 Above –20·0
ISOHYET
NOT IMPORTANT
DATA NOT AVAILABLE
LAKSHADWEEP (INDIA)
ANDAMAN AND NICOBAR ISLANDS (INDIA)

FIG. 12

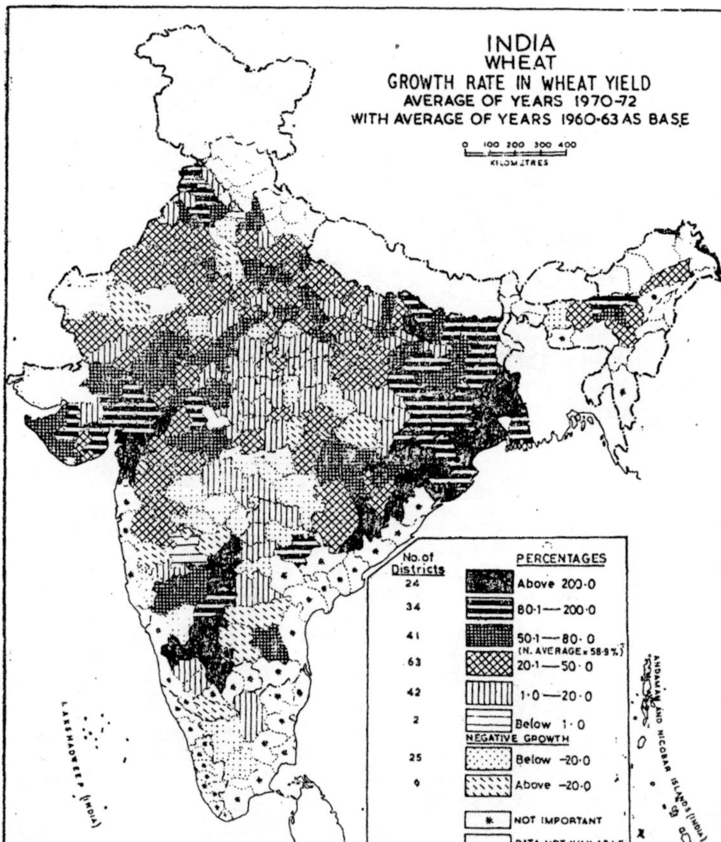
INDIA
WHEAT
GROWTH RATE IN WHEAT YIELD
AVERAGE OF YEARS 1970-72
WITH AVERAGE OF YEARS 1960-63 AS BASE
0 100 200 300 400
KILOMETRES
No. of Districts
PERCENTAGES
24 Above 200·0
34 80·1 — 200·0
41 50·1 — 80·0
(N. AVERAGE = 58·9%)
63 20·1 — 50·0
42 1·0 — 20·0
2 Below 1·0
NEGATIVE GROWTH
25 Below -20·0
0 Above -20·0
NOT IMPORTANT
DATA NOT AVAILABLE
LAKSHADWEEP (INDIA)
ANDAMAN AND NICOBAR ISLANDS (INDIA)

FIG. 13

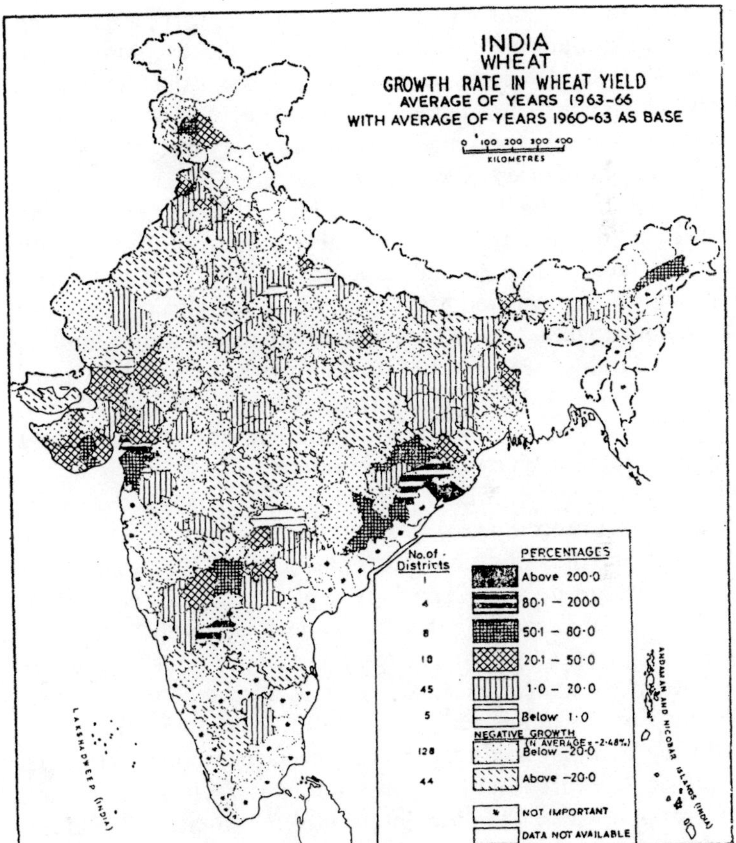
INDIA
WHEAT
GROWTH RATE IN WHEAT YIELD
AVERAGE OF YEARS 1963-66
WITH AVERAGE OF YEARS 1960-63 AS BASE
0 100 200 300 400
KILOMETRES
No. of Districts
PERCENTAGES
1
Above 200·0
4
80·1 – 200·0
8
50·1 – 80·0
10
20·1 – 50·0
45
1·0 – 20·0
5
Below 1·0
NEGATIVE GROWTH
(IN AVERAGE = -2·48%)
128
Below –20·0
44
Above –20·0
NOT IMPORTANT
DATA NOT AVAILABLE
LAKSHADWEEP (INDIA)
ANDAMAN AND NICOBAR ISLANDS (INDIA)

FIG. 14

analysis is attempted in the following pages. For this purpose, the districts are identified and aggregated on a regional basis. (Refer to Appendix-B in this chapter for districts grouped in each region). The regional aggregates of area under rice and wheat and production of rice and wheat are then set against the all-India totals of area and production of rice and wheat for the two blocks, viz., 1960-63 and 1970-72. Such a study at two points of time would bring out the temporal variations, in the relative contributions of different rice and wheat regions to the country's total. Regional average yields are calculated with a view to bring out inter-regional and intra-regional variations in the yield of these crops even within what can be considered as the core areas of rice and wheat concentration.

The districts have been grouped into different regions on the basis of similarity in natural environment taking into account factors like rainfall, topography, soil types and so on. The analysis is presented first region by region and then for all the regions together. Such an analysis helps in understanding yield variations, area and production between regions in terms of the difference in agro-climatic conditions of different regions. Certain land types favour the cultivation of rice whereas wheat requires different types of soil conditions. The core areas of wheat and rice (i.e., where area under these crops is over 10% of the gross cropped area) are found to be almost mutually exclusive as the conditions of climate and soil requirements of these crops are quite different. Exceptions to this are some districts in the Lower Ganges Plain and Eastern U.P. In order to assess intra-regional and inter-regional variations, it is necessary to analyse the role of national and human factors. In fact the landuse is influenced by physical, economic and institutional factors.[1] In India, there are sharp regional variations in physical features and they are dominant factors in shaping the cropping pattern in different parts of the country. Therefore, in order to analyse the crop production potential and to lay down priorities for certain regions and crops, it becomes necessary to have a regional framework adapted to administrative boundaries. It is important to identify and analyse regional physical complexes of evaluation of regional potentialities in crop production. The physical complexes indicate the priority to certain areas in respect of the allocation of the crop production target at any point of time.

The important elements of the natural environment, namely, climatic and its seasonal and regional variations, rock types affecting landforms and important topographical features and soils are used in identifying major regions and sub-regions. Using this regional framework an attempt is made in the following pages to analyse the regional and inter-regional patterns of rice and wheat and their relative share to the national total.

Rice regions

West Coast (*Region-1*)

This region consists of : (a) Coastal Maharashtra, (b) Coastal Karnataka, and (c) Malabar Coast.

This region accounts for 3.91 per cent of national area under rice during 1960-63 and 4.09 per cent in 1970-72. Its contribution to national production was 5.24 per cent in 1960-63 and 4.93 per cent in 1970-72.

In terms of average yield per hectare in kilograms, this region occupied the highest place with yield of 1339.53 kilograms per hectare in 1960-63 and 1465.9 kilograms per hectare in 1970-72.

Within the region the average area concentration was 50.47 per cent in 1960-63 and 56.23 per cent in 1970-72, that is, an increase of 5.76 per cent between 1960-63 and 1970-72.

Cropped landuse in this region is characterized by intensive cereal farming with emphasis on paddy and coconut cultivation. Paddy lands are mostly unirrigated as the rainfall is above the normal requirement of paddy (80″ and above) from June to September. During the dry winter, pulses are grown in these lands except that a very small per cent of lands are brought under paddy with the help of tank irrigation in Kerala coast and midlands. Recently groundnut has replaced pulses. The physical characteristics on the whole have a dominant influence on the existing pattern of landuse.

East Coast (Region-2)

This consists of : (a) Southern Coast of Tamil Nadu, (b) Central Andhra, and (c) Northern Orissa. This region stands out in contrast to Eastern Ghats region. It receives an annual rainfall of 40″-50″. There are significant variations in rainfall, landform and soil types in this region. The northern part, i.e., Orissa coast and the adjoining non-deltaic area of coastal Andhra receive rainfall mainly from south-west monsoons. Towards the south, the annual average is less than the north, but winter rains are important.

Crop cultivation, particularly of paddy and sugarcane depends partly on irrigation. There is a seasonal rhythm of rainfall which has introduced considerable variation, in the seasons of cropping. Paddy is the principal crop in coastal plains. There is also diversification in cropped landuse due to canal irrigation in deltas and tank irrigation elsewhere. Deltas have landuse with paddy, sugarcane and tobacco as in the district of Guntur. In the southern non-deltaic areas, there are contrasts in topography and climate as between the coastal area of Chingleput and South Arcot and Ramanathapuram. The Ramnad coast is dry and sandy and hence limits land productivity.

This region contributed 15.19 per cent to national area under rice in 1960-63 and 14.80 per cent in 1970-72. This region's share in national production is 18.93 per cent in 1960-63 and 18.33 per cent in 1970-72. In its average yield the region is next to West Coast with an average yield of 1244.8 kilograms in 1960-63 and 1416.8 kilograms per hectare in 1970-72. Within the regions the area concentration was 55.48 per cent in 1960-63 and 58.22 per cent in 1970-72. The area concentration in the region increased by 2.70 per cent between 1960-63 and 1970-72. Most of the areas in the region devote around 30 per cent of the gross cropped area to rice. This region contributed the highest percentage to national production in both 1960-63 and 1970-72. It is obvious, for, in this region are parts of Tamil Nadu and Andhra which are regarded as the rice bowl of India.

Peninsular Plateau (Region-3)

This region consists of : (a) Baghelkhand Plateau, (b) Chotanagpur Plateau, (c) Sambalpur area (Chattisgarh), (d) Orissa Hills, and (e) Bastar Hills. This region like the West Coast region or the East Coast region has its own features marked by humid climate (annual average over 70″ rainfall). Roughly this area also coincides with major resource region in which iron-ore, coal, mica, limestone, etc., and forests are related to the geological structure, landform and climate. Among the sub-regions, the Chattisgarh-Sambalpur area is of agricultural importance, particularly in cereal farming. The Chotanagpur Plateau is relatively more advanced. Baghelkhand Plateau has varieties. It is relatively of good quality from agricultural point of view.

This region occupied the first rank according to its share in total area under rice at the all-India level with 22.02 per cent in 1960-63 and 21.08 per cent in 1970-72. This is partly due to the large size of the region having twenty three districts. The region contributed 17.90 per cent to national production in 1960-63 and 15.48 per cent in 1970-72. In terms of average yield, the region's average is only 872 kg per hectare in 1960-63 and 863 kg per hectare in 1970-72. In area concentration the region had an average of 49.88 per cent in 1960-63 and 50.73 per cent in 1970-72.

North Eastern Valley Plains (Region-4)

This region consists of : (a) Brahmaputra-Tista Valley, and (b) Cachar Low Lands. In this region, climate is of moderate monsoonal type with hot season being shorter, low average temperature and high rainfall. In the valley floor, cereal farming is distinct.

The North Eastern Valley region's contribution to the total area under rice in the country was 6.09 per cent in 1960-63 and 6.33 per cent in 1970-72. The region's contribution to the country's production was 5.6 per cent in 1960-63 and 5.7 per cent in 1970-72. The average yield of rice was 920 kilograms per hectare in 1960-63 and it increased to 1022 kilograms in 1970-72. The area concentration within the region increased from

58.57 per cent in 1960-63 to 88.88 per cent in 1970-72, that is an increase of 30.31 per cent. The region has the highest increase in area concentration between 1960-63 and 1970-72. The reason for this higher average in 1970-72 may be due to the non-inclusion of two districts Jalpaiguri and Cooch Behar in this region due to the non-availability of reliable data on gross cropped area during 1970-72. It will be noted in Figure 1 that out of the nine districts four have more than 70 per cent area concentration. But in Figure 4, out of seven districts four have more than 70 per cent area concentration.

Lower Ganges Plain (*Region-5*)

The plain is transitional between the relatively dry *doaba* of the upper Ganges Plain and humid khadar of the Ganges delta. The soil is very fertile. The soil composition favours the cultivation of almost every crop that can be produced in semi-tropical low lands, but low in nitrogen and organic matter.

The Lower Ganges Plain region contributed 11.18 per cent in 1960-63 and 23.50 per cent in 1970-72. Its share in national production was 11.47 per cent in 1960-63 and 11.17 per cent in 1970-72. The average yield per hectare decreased from 941 kilograms per hectare in 1960-63 to 936 kilograms in 1970-72. In this region, area concentration increased from 39.76 per cent to 51.21 per cent, that is an increase of 11.45 per cent. This is the highest increase after the region of North Eastern Plains.

Ganges Delta (*Region-6*)

This region has alluvial soil and is a region of dying rivers and stagnant pools of water. This is a portion of the Lower Ganges Plain region.

The Ganges Delta region contributed 9.50 per cent to national area during 1960-63 and 9.20 per cent during 1970-72. Its contribution to national production was 11.77 per cent in 1960-63 and 11.53 per cent in 1970-72. The average yield increased from 1134 kilograms per hectare in 1960-63 to 1317 kilograms in 1970-72. The area concentration of rice was 73 per cent in 1960-63.[3]

Upper Ganges Plain—Eastern Uttar Pradesh (*Region-7*)

This region is transitional between the Punjab plain and Lower Ganges Plain. Rainfall (30″-50″) is mostly from south west Monsoon. But to the west, the region gets winter rains which is favourable to irrigated wheat. Soil is very fertile and favours multiple cropping.

The region of Upper Ganges Plain contributed 2.3 per cent to national area in 1960-63 and 1.47 per cent in 1970-72. Its share in national area under rice is the lowest after Northern Mountains. However, its percentage share in national production has increased from 1.61 per cent in 1960-63 to 4.93 per cent in 1970-72. There was an increase in area concentration from 34.40 per cent in 1960-63 to 37.17 per cent in 1970-72. This is the only region whose percentage share in national production increased by 3.32 per cent between 1960-63 and 1970-72.

Northern Mountains Region (*Region-8*)

This region consists of Jammu and Kashmir Hills. Agriculture is based on cereal farming. Sheep rearing is important. Soil in the valley floor is alluvial.

The region of Northern Mountains contributed the lowest percentage to national area, namely, 0.5 per cent in 1960-63 and 0.65 per cent in 1970-72.

4.5 REGIONAL PATTERNS AND CONCENTRATION OF RICE IN INDIA

For all the eight rice regions together, their share in national area under rice was 71.7 per cent in 1960-63 and 71.12 per cent in 1970-72. Although, there are inter-regional and intra-regional variations in area under rice to total cultivated area, the overall spatial pattern does not seem to have changed. The share of rice regions to total production of rice in the country between 1960-63 and 1970-72 has not changed. It was 73.20 per cent in 1960-63 and 73.14 per cent in 1970-72. With the exception of Punjab-Haryana area of the Indo-Gangetic Plain 8 regions comprising 93 districts out of 291 districts in the country constitute the core area where the strategy for modernisation of rice cultivation and growth of rice output should aim at regional specialisation in agriculture as in wheat growing areas of Punjab-Haryana. Intra-regional variations in yield are suggestive of the potential for increasing rice production. For example yields in the Upper and Lower Ganges Plain (Bihar and Eastern U.P.) Which are predominantly rice growing areas, have low yields in the range of 720 kg to 936 kg/hectare in contrast to the coastal plains where yields are in the range of 1400 kg-1500 kg/ha. This analysis adds realism to what is otherwise concealed in inter-State analysis by the use of national aggregates. Similar observation holds good for wheat as elaborated below.

Wheat Regions

Upper Ganges Plains (*Region-1*)

This region consists of Western U.P. plains, eastern U.P. plains and Bundelkhand tract. Climatically this region is somewhat transitional between the Punjab plains and Lower Ganges Plain. Rainfall is between 30″-50″. This region corresponds to region-7 in rice and has other similar characteristics.

The Upper Ganges Plains contributed 21.58 per cent to the country's area under wheat in 1960-63 and 29.60 per cent in 1970-72, an increase of 8.02 per cent. This increase is the highest among all the regions. The region's share to total production in the country was 25.59 per cent in 1960-63 and 25.16 per cent in 1970-72. In yields the region ranks second (1028 kg/ha) with Punjab plains occupying the first place (1098 kg/ha).

In terms of regional cropping pattern (i.e., area under wheat to total cropped area of the region), there has been an increase from 17.4 per cent in 1960-63 to 27.9 per cent in 1970-72, i.e., an increase of 10.5 per cent.

Lower Ganges Plains (*Region-2*)

This corresponds to Region-5 in rice and has similar characteristics in cropping pattern in that rice and wheat are complementary elements of agriculture.[5] This region's contribution to total area under wheat was 2.20 per cent in 1960-63 and 7.41 per cent in 1970-72 and in total production its shares during the same period were 1.92 per cent and 8.06 per cent respectively. It is thus evident that both in terms of area and production there is a distinct trend towards spatial spread of wheat cultivation in the adjoining regions where conditions for growing wheat are similar as in the Upper Ganges Plain. In terms of average yield the region has the lowest rank among all the other wheat regions. In 1970-72, some districts in the region had wheat cultivation where it occupied less than 10 per cent of gross cropped area. Such regions which at the moment appear peripheral but contiguous might emerge as core areas if extension of wheat cultivation in West Bengal anew is any indication of the spatial affinity of wheat.

Northern Mountains and Hill Region (Region-3)

This corresponds to Region-8 in rice. These are overlapping areas of rice and wheat.

The region's share to total wheat area in the country was 1.15 per cent in 1960-63 and 1.14 per cent in 1970-72. Its share in national production was only 0.96 per cent in 1960-63 and 0.34 per cent in 1970-72. In regional cropping pattern, however, wheat occupies about 46 per cent of the cropped area.

The average yield was 700 kilograms per hectare in 1960-63, which is almost half that of the Punjab Plains. In regional cropping pattern the share of wheat area was only 6 per cent of the total cropped area in 1960-63 and it had hardly changed over the years (6.9% in 1970-72).

Punjab Plain (Region-4)

This consists of : (a) Siwalik Foothills, (b) Central Irrigated Plains, and (c) the Southern Arid Margin. Except for Siwalik Foothills, the terrain is an alluvial tract, and margin has characteristics of Rajasthan except around Ganganagar which presents features similar to those of Punjab Plains as a result of extension of canal irrigation.

The Punjab Plains which include the central irrigated area of Haryana, accounted for 14.00 per cent of total area under wheat in 1960-63 and 19.99 per cent in 1970-72, an increase of 5.99 per cent. Its share in country's wheat production was 21.68 per cent in 1960-63 and 28.38 per cent in 1970-72, an increase of 6.70 per cent. The region has the highest average yield among all the other wheat region. (1898 kg/ha).

In 1960-63, it had an average yield of 1298 kilograms per hectare and it increased to 1898 kilograms per hectare in 1970-72. Incidentally, this is the region where a major break-through in wheat cultivation in terms of adoption of HYV and modern techniques of cultivation was witnessed as early as in 1965-66. In terms of cropping pattern the area under wheat increased from 19.70 per cent of the gross cropped area in 1960-63 to 24.64 per cent in 1970-72.

North Eastern Peninsular Plateau (Region-5)

This region consists of: (a) Rewa Plateau, (b) Bilaspur in Chattisgarh, (c) Baghelkhand Plateau, (d) Chambal Ravines, and (e) Bundelkhand Plateau. Rewa and Baghelkhand Plateau correspond to Region-3 in rice region and have similar characteristics. Chambal Ravines are marked by intense gully erosion and give rise to bad land topography. Bundelkhand plateau has 40″ rainfall and soil is of mixed red and black varieties and is potentially good for agriculture.

The share of this region to total area under wheat in the country was 8.18 per cent in 1960-63 and 8.79 per cent in 1970-72. Its share in total production was 7.06 per cent in 1960-63 and it declined to 4.76 per cent in 1970-72. The average yield per hectare was 830 kg/ha and remained almost the same between 1960-63 and 1970-72. Within the region, in terms of the cropping pattern wheat occupied 18.3 per cent gross cropped area in 1960-63 and 23.92 per cent in 1970-72.

Northern Peninsular Plateau (Region-6)

This region includes: (a) Malwa Plateau, (b) Narmada Trough, (c) Mahadeo Hills, and (d) Vidarbha. This region is also in the transitional zone of rainfall as between the semi-arid Rajasthan and humid north-eastern part of plateau. Vidarbha is an area of physical homogeneity. Narmada Trough has alluvial soils.

The North Peninsular Plateau region contributed 18.84 per cent to total area under wheat in the country in 1960-63 and 16.46 per cent in 1970-72. Thus, there was a decline in the relative share possibly due to variations in availability of irrigation unlike in the northern alluvial plains. Its share in country's production was 12.69 per cent in 1960-63 and 7.58 per cent in 1970-72, which also shows a decline possibly due to shift in acreage in preference to other crops specially when we find that the average yield increased from 560 kilograms per hectare in 1960-63 to 715 kilograms per hectare in 1970-72.

In cropping pattern wheat occupied 30.80 per cent of gross cropped area in 1960-63 and 30.68 per cent in 1970-72.

North-Western Transitional Region (*Region-7*)

This region consists of area around : (a) rest of the hill area of Mount Abu, (b) Aravalli, and (c) Mewar Plateau. The region gets rainfall between 15″-30″.

The region's share to total area under wheat in the country was 3.32 per cent in 1960-63 and 3.10 per cent in 1970-72. Its contribution to the country's production was 3.34 per cent in 1960-63 and 2.51 per cent in 1970-72. There was a sharp increase in yield per hectare of wheat during the same period from 847 kilograms per hectare to 1255 kilograms per hectare (Figure 12). The yield increase was the highest after the Punjab region. This was mainly due to the extension of irrigation facilities to this region after 1965.

In the regional cropping pattern wheat occupied 18.16 per cent of the gross cropped area[6] in 1960-63.

Apart from these major wheat growing regions there emerge isolated pockets of wheat concentration in the Central Peninsular Plateau and adjoining areas of the Western Ghats region. Because of the small size of these areas their contribution to national total area and production is insignificant.

4.6 REGIONAL PATTERNS AND CONCENTRATION OF WHEAT IN INDIA

Taking all the wheat regions together, in 1960-63, their share to total area under wheat was 70.06 per cent, and it increased to 87.75 per cent in 1970-72. This shows a strange tendency towards shift in cultivation towards wheat, a sign of strengthening of the core area to achieve regional specialisation unlike rice.

The share of all the wheat regions to national production was 73.46 per cent in 1960-63 and 77.08 per cent in 1970-72. This also corroborates the above finding. Inter-regional differences in yields are, however, quite marked and it should be the basis for formulating a spatial strategy for wheat cultivation complementary to rice and taking into consideration the crop combinations peculiar to agro-climatic conditions in the core areas of rice and wheat.

4.7 CORE AND PERIPHERIES OF RICE AND WHEAT REGIONS —A SUMMARY OF THE MAJOR FINDINGS

Rice and wheat are mutually exclusive except in parts of Madhya Pradesh, Jammu and Kashmir and the lower plains both in terms of cropping pattern and relative share to total area and production in the country and yields as between core areas of rice and wheat.

Prior to 1966, yield levels were very low for both the crops at the regional and national levels. But the increase in wheat yield was much sharper than the increase in rice yields after 1966. The national average yield of wheat increased very sharply, in 1970-72 over the earlier periods. Whareas, for the same period the increase in average yield of rice was not so sharp. This clearly shows that high yielding varieties have penetrated deeper within and between wheat regions and have been responsible for high wheat yields. However, post-1966 period shows better yields for both crops.[7]

Another contrasting feature to note is that while rice yields have increased very significantly in areas which are largely peripheral and have smaller area under cultivation of rice, wheat yields have increased very significantly, in both traditional areas as well as in areas brought anew under wheat cultivation. This raises an important question, namely, why have rice yields not increased that significantly in areas of very high concentration, which forms the basis of field investigation in this study.

The overall patterns of rice and wheat in the agricultural landscape show marked contrasts. The variations during 1960-63—1970-72 in respect of area, production and yield are more marked for wheat than for rice as is evident in the analysis of spatial and temporal variation in area concentration and growth rates of rice and wheat.[7] Rice regions have maintained their earlier position. For example, the percentage share of all the wheat regions to national area increased by 17.69 per cent between 1960-63 and 1970·72. But in the case of all the rice regions, the contribution remained the same. It was 71.72 per cent in 1960-63 and in 1970-72 it was 71.12 per cent. Similarly, the contribution of wheat regions to the country's production increased by 4.38 per cent between 1960-63 to 1970-72. Here again, the rice regions maintained their earlier position. In the case of Major Rice/wheat regions, yields have increased between 1960-63 to 1970-72, but the growth rates are different. The gap in yields within wheat regions seems to have narrowed while it has widened in rice region.

These findings and the regional framework provided the basis for the measurement of association among factors contributing to the spatial concentration and spread of rice and wheat and variations in their responses to high yielding variety as elaborated in the following chapters.

REFERENCES

1. L.S. Bhat, *et al.*, (1967), 'Regional Variations in National Environment and the Identification of Regional Physical Complexes for Agricultural Landuse in India. Indian Statistical Institute, New Delhi.
2. Data for 1970-72 was not available.
3. This region has an area concentration of 51.2 per cent for rice.
4. Comparable data for 1970-72 were not available in respect of gross cropped area.
5. This is corroborated by the findings from sample survey of landholdings in Chapter 4.
6. Refer Figures 1 to 8 and Chapter 4 for details.

APPENDIX A

CORE AREAS OF RICE CULTIVATION

Region 1	*West Coast*	*Districts*
	(i) Coastal Maharashtra	Thana, Bombay, Colaba and Ratnagiri
	(ii) Coastal Karnataka	North Kanara, South Kanara
	(iii) Coastal Kerala	Palghat, Trichur, Ernakulam, Alleppey (Malabar Coast), and Kanyakumari (Tamil Nadu)
Region 2	*East Coast*	
	(i) Southern	Ramanathapuram, Tanjore, South Arcot, North Arcot, and Chingleput (Tamil Nadu)
	(ii) Central	Nellore, Guntur, Krishna, West Godavari, Srikakulam, Visakhapatnam (Andhra)
	(iii) Northern	Ganjam, Balassore, Puri, Cuttack (Orissa)
Region 3	*North Eastern Peninsular Plateau*	
	(i) Baghelkhand Plateau	Shahdol, Surguja, Raigarh, Bilaspur, Durg, Raipur, Bhandara and Palaghat
	(ii) Sambalpur Area (Chattisgarh)	Kalahandi, Bolangir and Sambalpur
	(iii) Chotanagpur Plateau	Ranchi, Singhbhum, Hazaribagh, Dhanbad and Purulia
	(iv) Orissa Hills	Surendergarh, Dhenkanal, Mayurbhanj, Keonjhar
	(v) Eastern Hills	Koraput, Bastar and Chanda

Region 4	*North Eastern Valley*	*Districts*
	(i) Brahmaputra Tista Valley	Lakhimpur, Sivsagar, Nowgong, Darrang, Kamrup, Goalpara, Coochbehar, Jalpaigudi
	(ii) Cachar Low Lands	Cachar
Region 5	*Lower Ganges Plain*	
		Gorakhpur, Deoria, Azamgarh, Ghazipur, Champaran, Muzaffarpur, Darbhanga, Saharsa, Purnea, Bhagalpur, Santhal Parganas, Patna, Gaya, Shahbad, Malda and W. Dinajpur
Region 6	*Ganges Delta*	Murshidabad, Birbhum, Nadia, Burdwan, Bankura, Midnapur, 24 Parganas, Howrah and Hooghly
Region 7	*Upper Ganges Plains*	*Eastern U.P. Plains* : Gonda, Basti, Faizabad, Sultanpur
Region 8	*Northern Mountains*	*Jammu & Kashmir Hills* :
	(i) Kashmir Valley	Jammu, Srinagar, Baramullah and Anantnag
	Total Districts	93

CORE AREAS OF WHEAT CULTIVATION

Region 1	*Upper Ganges Region*	*Districts*
	(i) Western U.P. Plains	Bulandshahr, Meerut, Saharanpur, Muzaffarnagar, Bijnor, Moradabad, Kheri, Sitapur, Hardoi, Shahjahanpur, Bareilly, Badayun, Aligarh, Mathura, Agra, Etah, Mainpuri, Farukhabad, Etawah, Kanpur, Unnao, Lucknow, Bharatpur and Rampur
	(ii) Eastern U.P. Plains	Behraich, Bastar, Rae Bareily, Barabanki, Jaunpur, Varanasi, Allahabad, Fatehpur, Faizabad
	(iii) Bundelkhand Tract	Jalaun, Hamirpur, and Banda
Region 2	*Lower Ganges Plain*	Gorakhpur, Deoria, Ghazipur, Bhagalpur, Champaran, Saran, Shahbad, Patna, Gaya, Darbhanga, Saharsa, and Monghyr
Region 3	*Northern Mountains*	
	(i) U.P. Hills	Nainital
	(ii) J. & K. Hills	Poonch, Jammu and Udhampur
Region 4	*Punjab Plains*	
	(i) Siwalik Foot Hills	Gurdaspur, Amritsar, Ludhiana, Jullundur, Hoshiarpur, Patiala, and Ambala
	(ii) Central Irrigated Area	Ferozpur, Bhatinda, Sangrur, Karnal, Rohtak, Hissar, Gurgaon, and Ganganagar

Region 5	*North Eastern Peninsular Plateau and Part of North Peninsular Plateau*	*Districts*
	(i) North Eastern Portion	Panna, Satna, Rewa, Mirzapur, Jabalpur, Mandla and Bilaspur (All from M.P. except Mirzapur—U.P.)
	(ii) Northern Portion	Chambal ravines, Morena (MP), Bhind (MP), Gwalior (MP), Datia (MP), Shivpuri (MP), and Sawai Madhopur (Rajasthan)
	(iii) Bundelkhand Plateau	Jhansi, Tikamgarh and Chattarpur
Region 6	*Peninsular Plateau*	
	(i) Malwa Plateau	Mandasaur, Ratlam, Ujjain, Indore, Dewas, Shahjapur, Guna, Rajgarh, Vidisha, Sahore, Raisen, Sagar, Damoh, Kota and Jhalawar
Region 7	*North-Western Transitional Range*	
	(i) Aravalli Hills	Pali
	(ii) Mount Abu	Sirohi, Udaipur
	(iii) Mewar Plateau	Tonk, Bhilwara, Chittorgarh and Bundi (Dungarpur belonging to Gujarat uplands included here for statistical convenience)
Others	(i) Northern-Western Ghats	Nasik
	(ii) Central Peninsular Plateau	Dharwar
	Total Districts	114

5

Regional Variations in Rice and Wheat Responses

An Analysis of Factors Influencing Growth in Area and Yield

5.1 INTRODUCTION

IN Chapter 4, spatial concentration and growth rates in yields of rice and wheat have been analysed first for all the rice and wheat districts and then for core regions of rice and wheat, which account for 71.00 per cent of the national area and 73.14 per cent of national production in the case of rice and the core wheat regions account for 87.75 per cent of national area and contribute 77.08 per cent to national production.[1] In the case of regions where area concentration is high as in rice, it was noticed that yield growth rates were not so high as in regions like Punjab-Haryana where rice occupies relatively a small percentage of cultivated area. In the case of wheat, it was noticed that the North-Western transitional region (Rajasthan) had shown highest growth rates in yield particularly between 1963-66 and 1970-72. This was the result of extension of irrigation facilities in the region. These inter-regional variations in area and yield are influenced by a number of explanatory variables and by examining these relationships, we can find useful clues to substantiate inter-regional variations in responses to high yielding variety, fertilizers and so on.[2]

5.2 REGIONAL VARIATIONS IN AREA AND YIELD RESPONSES

This chapter attempts to examine inter-regional variations in the responses of growth rate in area and yield per hectare of rice and wheat for the core regions of rice and wheat.[3] This has been analysed at three points of time with the help of stepwise regression. These three points of time refer to the following :

(i) Growth rate in average area 1970-72, with the average of 1960-63 taken as base.
(ii) Growth rate in area during 1970-72 with the average of 1963-66 taken as base.
(iii) Growth rate in area during 1963-66 with the average of 1960-63 as base.

The idea in taking the period 1970-72, with two different base averages, was that it could bring out the contrasts as between modernisation of agriculture initiated after 1966 and the traditional agricultural practices prevailing prior to that date. In this model, growth rate in area is the dependent variable and is explained in terms of the following explanatory variables.

X_1 Growth rate in gross cropped area.
X_2 Percentage of gross irrigated area to total cropped area.
X_3 Percentage of gross irrigated area of rice and wheat to area under rice and wheat.
X_4 Growth rate in yield per hectare (in rupees).
X_5 Growth rate in yield per hectare (in kilograms).

Yield per hectare is a function of many factor inputs. Some of these factors are controllable in the sense that the quantity of some of the factor inputs like, irrigation, availability of human labour, agricultural practices and so on, can be controlled by the farmer. Then there are uncontrollable factors like weather which is important. In fact, any study of variations affecting production and yield is beset with basic limitation if it is undertaken without taking into account, weather-crop relationships. Such a study would help in separating out the effects of weather on production from that of developmental efforts on the crop. But in practice, in carrying out an inter-regional analysis based on districtwise data especially in a country like India with marked regional contrasts, one has to be content with an analysis which embodies the major element of weather.

Apart from weather, there are economic and institutional considerations, which also affect the yield. Availability of credit at the right time and rate of interest paid and related factors also indirectly affect yields. To get these types of information, one has to fall back on field data collected through sample survey of households.[4]

However, in the present chapter, an attempt is made to analyse the inter-regional variations in yield per hectare of rice and wheat in 1967-68 (time block I) and 1970-71 (time block II) in terms of certain explanatory variables with the help of step-wise regression.

In the model, the dependent variable is yield per hectare (in kilograms) in 1967-68 and 1970-71. The explanatory variables are :

X_1 Fertilizer consumption in kilograms per hectare.

X_2 Number of tractors per thousand hectares.

X_3 Number of pump sets and oil engines per thousand hectares.

X_4 Area irrigated of rice/wheat as a percentage of area under rice/wheat.

X_5 Area under high yielding variety of rice/wheat as a percentage of area under rice/wheat.

X_6 Price of rice/wheat in rupees per kilogram.

5.3 RICE AREA RESPONSES

Stepwise Regression Results

The results are presented first region by region followed by results for all regions combined. Tables 5.1 to 5.7 relate to the area responses of rice regions to the variables selected for analysis.*

*Refer Appendix-B for regression tables at the end of this chapter.

West Coast Region

Table 5.1 : This relates to the West Coast Region. In 1970-72 with the average of 1963-66 as base, all the three variables, namely, area irrigated (X_2), percentage of irrigated area of the crop (X_3) and growth rate in yield (X_5) have negative influence on rice area, their T values are significant at 80 per cent levels. This is partly because rice is wholly rainfed during South-West Monsoon. In the final step of the equation, their T values become significant at 90 per cent level. The R bar square value has increased from 0.28 in step 1 of the equation to 0.71 in step 3 of the equation. When the base is shifted to 1963-66 then the results become insignificant at all levels. Most of the variables have negative regression co-efficients. In 1963-66 with the average of 1960-63 as base the results are mostly insignificant. A thing to be noted is that the R square values are also very high in all the three periods. The R bar square value has, however, dropped steadily after step 1 of the regression equation (Table 5.1). This area consists of parts of Malabar and south Konkan coast, and annual rainfall is very high (above 80″) in this region. Perhaps this alone is sufficient to explain variations in rice area.

East Coast Region

Table 5.2 : This brings out the regression results for the East Coast region. Yield per hectare (X_5) is negatively correlated with rice area. However, the T value is insignificant. After the introduction of gross cropped area (X_1) in equation 1.2 the R square value increases to 0.38 from 0.3 and after this the R bar square value declines. The R bar square value gives us the total explanatory power of the equation. Gross cropped area improves the explanatory power of the equation and is in fact most important among other factors in influencing area under rice in this region. This variable along with gross area irrigated explains 53 per cent of total variation in rice area in the region.

However, when the base is changed to 1963-66 gross cropped area explains 23 per cent of the total variation in the rice area and with gross irrigated area as another variable it explains 57 per cent of the total variation. The R bar square value has increased from 0.16 in equation 1.1 to 0.48 in equation 1.2. The T values are significant at 95 per cent level.

In 1963-66, with the average of 1960-63 as base, only yield (in rupees per hectare) emerges as a factor influencing rice area. It explains 24 per cent of the total variation on rice area. The T value is significant at 95 per cent level.

An interesting feature of this region is that the yield per hectare (in rupees) alongwith gross cropped area and gross area irrigated influenced rice area significantly.

North-Eastern Peninsular Plateau Region

Table 5.3 : In the North-Eastern Peninsular Plateau yield per hectare (in kilogram) influenced rice area in 1970-72 with the average of 1960-63 as base. It explains 37 per cent of the total variation in rice area. The result is significant at 95 per cent level. The area devoted to rice does not seem to be influenced by irrigation availability or the level of yield. This is because yields are low in comparison with other crops and where irrigation is available it is devoted to cash crops.

In 1963-66 with the average of 1960-63 as base once again yield (in kilograms per hectare) has a negative influence on rice area. Yield (in rupees per hectare) has a positive influence on rice area. The inclusion of this variable improves the R bar square value from 0.11 in equation 1.2 to 0.15 in equation 1.3, the result is insignificant. It is gross area irrigated that has a positive and significant influence on rice area. In the last step of the equation result is significant at 90 per cent level for this variable.

Assam Plains Region

Table 5.4 : In this region the variation in percentage of rice area irrigated has influenced area under rice in 1970-72 with the average of 1960-63 as base. The result is significant at 90 per cent level. In fact, this variable explains 64 per cent of the total variation in rice area. In 1970-72, with the average of 1963-66 as base, gross area irrigated has a negative influence on rice area. The result is significant at 90 per cent level.

The extent of area devoted to rice seems to be influenced

by the availability of irrigation and wherever irrigation is available it is used to grow rice during three seasons in a year (*aus*, the *aman* and *boro* varieties of rice). In this region the gross irrigated area had a negative influence on area under rice. In this region also it appears that it is the adequate rainfall that influences cultivation of rice and not the availability of irrigation.

Ganges Delta and Assam Plains

Table 5.5 : This relates to the region of Ganges Delta combined with the region of Assam Plains, yield per hectare (in kilograms) has a positive influence on rice area. The result is, however, significant only at 80 per cent level. With 1963-66 as base in 1970-72, yield per hectare influenced rice area negatively. None of the variables explain variation in rice area significantly. In 1963-66 with the average of 1960-63 as base, the results are not very significant. The gross cropped area is the only factor that has a positive influence on rice area. The R bar square value is negative in the last step of the equation.

Lower and Upper Ganges Region

Table 5.6 : In this region, in 1970-72 with the average of 1960-63 as base, yield growth rate (in rupees per hectare) has a positive and significant influence on rice area. The result is significant at 99 per cent level. In 1970-72, with 1963-66 average as base, the growth rate in gross cropped area has a negative influence on rice area. In 1963-66, with the average of 1960-63 as base, all the results are insignificant. Thus, in this region it is only in 1970-72 with the average of 1960-63 as base that the result is significant. Rice is not as important as other crops such as wheat and sugarcane and hence where conditions for intensive cultivation and expansion of cultivated area are favourable the additional area is devoted to crops other than rice.

All Rice Regions

Table 5.7 : When we consider all the rice regions together it is clear from Table 5.7 that the results do not corroborate the findings, for this is because of the contrasts that exist among different rice regions. In 1970-72 with the average of 1960-63 as base yield (in rupees per hectare) is significant at 99 per cent level in its influence on rice area However, it explains only 16 per cent of the variation in rice area.

In 1970-72 with the average of 1963-66 as base, gross area irrigated has a negative influence on rice area and the result is significant at 98 per cent level. In 1963-66 with average of 1960-63 as base, the results are mostly insignificant. This result is in contrast to the results for the wheat regions considered together. The variables selected explain variation in wheat area far more satisfactorily than for rice area. The reason for this may be that rice cultivation is closely associated with the availability of rainfall in the core regions of rice, whereas, wheat cultivation is highly correlated with irrigation availability.

5.4 VARIATIONS IN RICE YIELD RESPONSES

West Coast Region

Table 5.8 : In this region, in 1967-68, the results are not very satisfactory. Use of tractors influenced rice yield in this period and the result is significant at 90 per cent level. In 1970-72, once again the same variable has a significant influence on rice yield. The result is significant at 95 per cent level in equations 1.1 and 1.2 and at 98 per cent in equation 1.3. Percentage of area under high yielding variety has negative influence on rice yield in this period and this result is significant at 90 per cent level in equation 1.3. Both these variables together explain 88 per cent of total variation in rice yield, fertilizer consumption per hectare explains negatively rice yields in the region. In West Coast region the modern agricultural practices and use of inorganic input etc. is relatively recent.

East Coast Region

Table 5.9 : In contrast to the West Coast region, in the East Coast during 1967-68 fertilizer consumption per hectare is significant in its influence on rice yield. During 1970-71 fertilizer consumption per hectare and proportion of area under high yielding variety are quite significant in explaining the increase in yield of rice. These two variables together explain 79 per cent of rice yield. This holds good because the region includes Tanjore Delta—a rice bowl of Tamil Nadu and Krishna Godavari Delta of Andhra Pradesh and the Mahanadi Delta of Orissa Coast.

North-Eastern Peninsular Plateau Region

Table 5.10 : In this region, both during 1967-68 and 1970-71 proportion of rice area irrigated has a positive and significant influence on rice yields. In 1970-71, price of rice per kg is also an important factor. In this year percentage of rice area irrigated and price together explains 63 per cent of rice yields in the region.

Assam Plains and Ganges Delta Region

Table 5.11 : In the region of Assam Plains and Ganges Delta (West Bengal) fertilizer consumption per hectare influenced rice yield during 1967-68 and 1970-71. In the latter year it has a positive and significant influence on yield and explains 58 per cent of the total explained variation in yield. In 1967-68 17 per cent of variations in yield was explained by this variable.

Lower and Upper Ganges Region

Table 5.12 : In the Lower and Upper Ganges region, during 1967-68 tractors per hectare and proportion of area under high yielding variety, however, had negative influence on rice yields. The value of the regression co-efficient was rather high. A significant feature to note is that during 1970-71 the values are negative for variables like tractors and proportion of area under high yielding variety. They negatively explain 20 per cent of variation in yield. This is the rice region where none of the explanatory variables are relevant in explaining variations in yield during 1970-71.

All Rice Regions

Table 5.13 : When all the rice regions are considered together during 1967-68, fertilizer consumption per hectare and proportion of rice area irrigated are positive and significant in their influence on rice and they explain 37 per cent of the total variation. The results are significant at 99 per cent level. However, inclusion of fertilizer consumption per hectare reduces the value of R bar square. But it has a significant value in itself (at 90 per cent level).

In rice yield on the whole fertilizer cousumption per hectare and proportion of rice area irrigated are the two variables which explain positively and significantly yield levels in most rice regions, both during 1967-68 and 1970-71. The same observation is true for all the rice regions considered together. So far as the total explained variation by all the variables is concerned during 1970-71 it is as high as 92 per cent in West Coast, 79 per cent in East Coast, 58 per cent in North Eastern Peninsular Plateau and 20 per cent in Lower and Upper Ganges region. It was above 60 per cent in the East Coast region and Lower and Upper Ganges region. It was 35 per cent in the West Coast region.

5.5 VARIATIONS IN WHEAT AREA RESPONSES

Tables 5.14 to 5.20 give the stepwise regression results first regions by region for the time blocks and then for all the wheat region taken together.

Upper Ganges Plain Region

Table 5.14 : This brings out that for the Upper Ganges Plain, during 1970-72, with 1960-63 average as base, growth rate in gross cropped area is important in its influence on growth rate in area under wheat. It has a T value which is significant at 90 per cent level. However, the explained variation as given by R square value is only 9 per cent when the base year is changed to 1963-66. Variations in percentage of gross area irrigated to gross cropped area has come out significant. The regression co-efficient gives a T value of 2.56 which is significant at 99 per cent level.

During 1963-66 with 1960-63 average as base, gross area irrigated as a percentage of area under wheat has a negative influence on growth rate in area and the T value is significant at 80 per cent level. In this region after 1963-66, area under wheat is influenced by gross area irrigated. With 1960-63 average as base, increase in total cropped area explains the variations in wheat area in 1970-72. Between 1960-63 and 1963-66 both these factors are not important.

Lower Ganges Plain Region

Table 5.15 : In 1970-72 with the average of 1960-63 as base, gross area irrigated as a percentage of gross cropped area has an important influence on growth rate in wheat area in this region. The T value is significant at 98 per cent level and this variable explains 58 per cent of total variation in wheat area, when the base is changed to 1963-66, in 1970-72 irrigated area of wheat as a percentage of total area under wheat has an important influence on wheat area. The value of regression co-efficient is highly significant for these variables. During 1963-66 with 1960-63 as base both gross cropped area (X_1) and yield per hectare in kgs significantly influenced area under wheat (equation 1.5).

Punjab Plains

Table 5.16 : This gives the regression results for the region of Punjab Plains. In this region, during 1970-72, with the average of 1960-63 as base, only gross area irrigated as a percentage of gross cropped area is significant in its influence on wheat area. The T value is significant at 80 per cent level. With the average of 1963-66 as base, in 1970-72, in this region significant results emerge. There is a negative relationship between yield per hectare (kgs) and growth rate in area. Irrigated wheat area is also negatively associated with area increases. Growth rate in gross area irrigated as a percentage of gross cropped area had a positive influence on wheat area increase. The inclusion of the variables at every step increased the R square and R bar square values, upto equation 1.3. Thereafter, the R bar square value decreased when growth rate in yield is introduced (equation 1.4, X_4) and it is further reduced when growth rate in gross cropped area is introduced in equation 1.5. All the variables together explain 39 per cent of total variation in wheat area. Growth rate in gross area irrigated is significant in all the equations (at 80% level) and growth rate in yield (in kgs) is also significant except in the 2nd equation (1.2) when gross area irrigated is introduced. This shows that in the region of Punjab Plains both gross area irrigated and percentage of gross cropped area and yield per hectare (in kgs) significantly influenced growth rate in area. While yields explain area growth rate negatively, irrigation explains it positively. Thus area under wheat is explained by increase in gross area irrigated in this time block. During the period 1963-66 with 1960-63 average as base, only one variable emerged and that is yield per hectare in kilogram. It had a positive and significant influence on growth rate in wheat area.

North Eastern Peninsular Plateau Region

Table 5.17 : This brings out that in the region of North Peninsular Plateau consisting of parts of Madhya Pradesh and the Southern periphery consisting of parts of Uttar Pradesh plains, in 1970-72 with the average of 1960-63 as base, all the results are significant. Growth rate in gross cropped area explains wheat area negatively. The T value is 2.38, which is significant at 95 per cent level. The other variable that is significant in explaining growth rate in wheat area is the extent of irrigated area under wheat. It has a positive influence on growth rate in wheat area. The inclusion of growth rate in gross area irrigated as a percentage of gross cropped area increased the T value for this variable. The total explained variations by both these variables is 57 per cent. Thus, irrigation is an important factor for increase in wheat area in this region. When the base is changed to 1963-66, in 1970-72, growth rate in gross cropped area has a significant influence on growth rate in wheat area and it explains 20 per cent of total variation in wheat area.

In 1963-66 with the average of 1960-63 as base, all the results are insignificant. Growth rate in gross cropped area, irrigated area under wheat and yield in value terms together explained only 7 per cent of total variation in wheat area. Thus in this region, the base 1960-63 with respect to 1970-72, is most relevant in explaining growth rate in wheat area in terms of the chosen variables.

North Peninsular Plateau

Table 5.18 : In this region, yields have nothing to do with any of the variables selected in this study. The R bar square is negative throughout. This brings out the fact that wheat cultivation is still traditional and denotes lack of any change in their methods.

North-Western Transitional Region

Table 5.19 : In the North-Western Transitional region consisting mainly of Rajasthan, most of the T values are insignificant in 1970-72 with the average of 1960-63 as base. However, due to the introduction of gross area irrigated as one of the variables, both in equations 1.2 and 1.3 in this table, the R square value has increased to 0.72 (72%) and 0.78 (78%) and in equation 1.4, it increased to 0.79 (79%). This variable, therefore, had a positive and significant influence on growth rate in wheat area in the regions. The R bar square value has however declined gradually, once other variables are introduced (Equations 1.3 and 1.4).

In 1970-72 (with 1963-66 average as base) gross cropped area and gross irrigated area are significant in their influence on growth rate in wheat area. Wheat area irrigated has a negative influence. This may be due to the fact that gross irrigated area itself is quite significant in its influence on wheat area. It will be noticed that in equation 1.2 when gross irrigated area is introduced alongwith gross cropped area, the R square value increases from 0.40 (40%) to 0.96 (96%). The R bar square values have also increased. All the variables together explain 98% of total variation in area. The T values for both gross cropped area and gross area irrigated are highly significant. This is obviously to be expected in a region where, prior to 1966 irrigation facilities were not available. After 1966 in

Rajasthan, introduction of irrigation has brought about an increase in area under wheat. This is corroborated by the table for the period 1963-66 with 1960-63 as base wherein the influence of gross irrigated area under wheat was negative. In this period average growth rate in gross cropped area is significant in its influence on growth in wheat area.

All Wheat Regions

Table 5.20 : For all the wheat regions together, with 1960-63 average as base, in 1970-72, all the variables together explain 28 per cent wheat area. All the variables except yield per hectare (in kgs) emerge with T values which are high. All the five independent variables are positively and significantly correlated with growth rate in wheat area. Of all the variables wheat area irrigated is most significantly correlated with wheat area. Out of the total R square value of 0.28 (28%) it explains 0.19 (19%).

5.6 VARIATIONS IN WHEAT YIELD RESPONSES

The results pertaining to yield responses of wheat are not very encouraging. Though variables selected here, explain wheat yields both in 1967-68 and 1970-71 when the results are considered region by region, yet, when all the wheat regions are considered together the variables are far more relevant in explaining wheat yields.

Upper Ganges Plain Region

Table 5.21 : This brings out that in the Upper Ganges Plains percentage of wheat area irrigated is the variable emerging as important and significant in its influence on wheat yields both in 1967-68 and 1970-71, explaining 25 per cent and 16 per cent of total variations in wheat yield respectively.

Lower Ganges Plain Region

Table 5.22 : In this region tractors and pumpsets combined with proportion of wheat area irrigated explain 65 per cent of wheat yield in 1967-68. However, percentage of wheat area irrigated is negatively correlated with wheat yields. In 1970-71, tractors are negatively correlated with wheat yields.

Punjab Plains

Table 5.23 : For the region of Punjab Plains, pumpsets and engines are important in their influence on wheat yield in 1967-68. This variable explains 44 per cent of total variation in wheat yield in 1970-71. Percentage of area under high yielding variety had insignificant result for Punjab region where after 1966 wheat yields have increased sharply.

North-Eastern Peninsular Plateau

Table 5.24 : Percentage of wheat area irrigated in North-Eastern Peninsular region both in 1967 and 1970-71 explains 53 per cent of total variation in 1967-68 and 50 per cent of total variation in 1970-71. The T values are highly significant in both time periods.

North Peninsular Plateau

Table 5.25 : In this region, yields do not seem to have been influenced by variables selected for study. The R bar square is negative throughout. This brings out the fact that wheat cultivation is still traditional and of subsistence level in contrast to Punjab-Haryana region.

North-Western Transitional Region

Table 5.26 : In this region, in 1967-68 pumpsets and oil engines have a significant influence on wheat yield. The influence of price is negative on wheat yields. However, both these variables together are relevant in explaining their association with wheat yields, as R bar square move up from 0.46 in equation 1.1 to 0.72 in equation 1.2. However, in 1970-71 percentage of wheat area irrigated has a positive and significant influence on wheat yield in the region. This is to be expected in a region covering dry area of Rajasthan, because provision of irrigation facilities has really improved yields. This is corroborated by the spatial analysis in Chapter 4.[5] The areas of Rajasthan have shown high increase in wheat yield. This variable has a significant T value. Both R square and R bar square values have increased for equation 1.1 to 1.2. The negative correlation between pumpsets and yield, and the positive influence of irrigation on wheat yield together explain 76 per cent of total variation in wheat yields.

All Wheat Regions

Table 5.27 : This gives the regression results of all the wheat regions taken together. In 1967-68 all the variables except percentage of area under high yielding variety appear in all the equations. However, only two variables, namely, pumpsets per thousand hectares and wheat area irrigated improve the R bar square value. After that all through, the T values are significant at 80 per cent level. The R bar square and R square values have remained the same.

In 1970-71, however, there is a marked change in the behaviour of variables. All the variables except price of wheat influence wheat yields significantly. The percentage of area irrigated of wheat has increased the R bar square from 0.54 to 0.66. The percentage of area under high yielding variety of wheat influenced yields significantly in 1970-71. On the whole fertilizer consumption per hectare, tractors (per thousand hectare), pumpsets, irrigated area of wheat and percentage of area under high yielding variety of wheat, together explain 69 per cent of total variation in wheat yield in step 1.4. However, with the inclusion of fertilizer consumption in equation 1.5 the explained variation is 65 per cent. The R bar square value also declined. This variable was negatively correlated with wheat yield in 1970-71. Thus, all the variables in 1970-71 have significant value except for X_1 (fertilizer consumption). This is a very significant result which reflects upon the homogeneity of the wheat growing regions.

5.7 MAIN RESULTS

In so far as the relative behaviour of variables is concerned they have behaved more uniformly in wheat than in rice regions, in 1967-68 as well as in 1970-71. In 1970-71 in all the wheat regions except Punjab plains, proportion of area irrigated is significant. In the case of rice regions, only in the region of North Eastern Peninsular Plateau this variable has emerged as a significant factor in explaining rice yield in the regions. Another significant difference between rice and wheat regions is in the behaviour of two variables, namely, fertilizer consumption per hectare and price per kg of the crop concerned. In all the rice regions combined, both these variables are positive and significant in their influence on yields where as both these are insignificant for wheat regions. This establishes the fact, that a higher price per kilogram of rice acts as an incentive for improving yields of rice and hence the production.

A third feature that stands out in a comparison between rice and wheat regions inputs is that farm equipment and machinery are actually used by very few farmers even in 1975-76 as was shown by the data collected through field survey. Very few households in the rice regions possess tractors, pumpsets and oil engines. In fact, these variables have not emerged as significant. Fertilizer consumption per hectare and proportion of area under high yielding variety is significant in the rice regions. But when all the rice regions are combined the proportion of rice area irrigated is significant.

So far as the relevance of the variables in explaining wheat yields is concerned during 1970-71 the total explained variations of the variables is highest in North Western Transitional region (Rajasthan) 76 per cent. This is followed by Punjab plains region (62%), North-Eastern Peninsular Plateau (50%) and Northern Peninsular Plateau 35 per cent. For all the wheat regions combined it is 69 per cent whereas for all the rice regions, it was only 40 per cent. For wheat regions, the variables explain 60 per cent of the total variation in wheat yield during 1967-68 and 1970-71.

This goes to show that the variables chosen are more relevant and useful in explaining wheat yields rather than rice yields. This is perhaps due to the fact that the requirements for rice cultivation are more difficult to meet. It requires particular types of soil and assured irrigation and water supply. Many rice regions do not cultivate the crop during the three seasons and it is mostly a Kharif crop except in areas where rice is cultivated with irrigation during the dry season. Thus, fluctuations in weather condition may be a more determining factor in explaining rice yields in the absence of assured irrigation facilities. The regression results for rice regions should be viewed from the angle of fluctuations in weather conditions.

The foregoing analysis is based on secondary data. However the analysis does bring out some important results. In order to see to what extent these results reflect ground reality one has to undertake detailed enquiry in different regions. After all, ultimately the yield variations within and between regions are influenced also by factors such as the size of holding, credit availability and interest rates on capital borrowed, etc. To throw light on the association between size of holding and other factors determining crop yields, a case study of 198 landholdings has been attempted in the following chapter, on the basis of information collected through sample survey.

REFERENCES

1. These figures represent average of 1970-72.
2. It is better to conduct such studies at district or below district level to get a regional perspective. The results obtained at these levels are more useful in the context of integrated area development.
3. Refer Appendix-A in Chapter 4 for core regions of rice/wheat.
4. This aspect forms part of the sample survey. Refer Chapter 6 for details.
5. Refer Figure 12 in Chapter 4.

APPENDIX-B

TABLE 5.1 Area Responses of Rice

Stepwise Regression Results
Dependent Variable : Growth in Area
No. of Independent Variables : 5
Relevant Number of Samples for Regression : 10
West Coast Region

Equation Number	Constant	Regression Co-efficient of					R^2	$\bar{R}^2$	Constant	T Value				
		X_1	X_2	X_3	X_4	X_5				X_1	X_2	X_3	X_4	X_5
For 1970-72 (1960-63 average as base)														
1.1	16.81 (5.13)		—1.38 (0.67)				0.37 (37%)	0.23	3.27		2.05[5]			
1.2	22.30 (5.09)		—1.5 (0.56)	—1.0 (0.50)			0.62 (62%)	0.49	4.41		2.66[5]	1.98[5]		
1.3	28.99 (4.50)		—1.55 (0.42)	—1.21 (0.39)		—0.38 (0.16)	0.82 (82%)	0.71	6.44		3.62[4]	3.08[4]		2.3[4]

TABLE 5.1 (*Contd.*)

Equation Number	Cons-tant	Regression Co-efficient of					R^2	$\bar{R}^2$	Constant	T Value				
		X_1	X_2	X_3	X_4	X_5				X_1	X_2	X_3	X_4	X_5
					For 1970-72 (1963-66 average as base)									
1.1	15.64 (6.99)			—2.20 (1.20)			0.32	0.22	2.61			1.83*		
1.2	19.92 (6.49)			—2.14 (1.13)		—0.35 (0.26)	0.48	0.31	3.06			1.89^{5}		1.37*
1.3	21.58 (6.03)		—0.73 (0.48)	—1.88 (1.04)		—0.45 (3.55)	0.64 (64%)	0.43	3.57		1.49*	1.80*		1.84*
1.4	76.21 (122.04)		—2.14 (1.56)	—3.24 (5.49)	—2.45 (2.57)	—2.93 (2.55)	0.71 (71%)	0.42	0.61		1.36*	0.59*	0.95*	0.82*
1.5	79.43 (128.41)	—.24 (0.29)	—2.38 (1.65)	—3.63 (5.75)	—2.62 (2.69)	—3.17 (3.71)	0.76 (76.4)	0.37	0.61	0.83*	1.44*	0.63*	0.97	0.85*

For 1963-66 (with 1960-63 average as base)

1.1	3.64 (1.32)	1.04 (0.59)			0.30	0.20	2.74	1.75*		
1.2	4.90 (1.19)	1.89 (0.56)	—0.67 (0.26)		0.66	0.54	4.09	0.36*	2.50[4]	
1.3	6.51 (1.70)	1.85 (0.49)	—0.45 (0.27)	—0.05 (0.03)	0.78	0.64	1.64	3.75[4]	1.65*	1.64*

TABLE 5.1 (*Contd.*)

Note : I. Figures in parenthesis denote the standard error

II.1. Significant at 99% level

2. Significant at 98% level

3. Significant at 95% level

4. Significant at 90% level

5. Significant at 80% level

* Insignificant

1, 2, 3, 4, 5, and * denote value of T, for given degrees of freedom at specified levels of significance

III. *List of Independent Variables*

X_1 Growth rate in gross cropped area

X_2 Variation in gross area irrigated of the cropped area

X_3 Variation in percentage of area irrigated of the crop to total area under crop

X_4 Growth rate in yield (Rupees/ha)

X_5 Growth rate in yield (Kgs/ha)

IV. Equation numbers 1.1, 1.2, etc., indicate the steps of regression

TABLE 5.2 **Area Responses of Rice**

Stepwise Regression Results
Dependent Variable : Growth Rate in Area
No. of Independent Variables : 5
Relevant Number of Samples for Regression : **12**
East Coast Region

Equation Number	Constant	Regression Co-efficient of					R^2	$\bar{R}^2$	Constant	T Value				
		X_1	X_2	X_3	X_4	X_5				X_1	X_2	X_3	X_4	X_5
1	2	3	4	5	6	7	8	9	10	11	12	13	14	15
For 1970-72 (1960-63 average as base)														
1.1	—0.86 (0.17)		0.57 (0.49)				0.12	0.03	0.17		1.17*			
1.2	—1.61 (1.17)	0.86 (0.33)	1.34 (0.48)				0.50	0.38	0.38	2.61[4]	2.75			
1.3	0.65 (5.49)	0.98 (0.37)	1.45 (0.52)			—0.12	0.53	0.35	0.11	2.59[4]	2.76[3]			0.17*
1.4	– 3.38 (10.09)	0.88 (0.35)	1.66 (0.30)		0.16 (0.10)	—0.53 (0.30)	0.65	0.45	0.33	2.51[4]	2.51[3]		1.55[5]	1.73[5]

TABLE 5.2 (*Contd.*)

1	2	3	4	5	6	7	8	9	10	11	12	13	14	15
						For 1970-72 (1963-66 average as base)								
1.1	1.45 (3.60)	0.37 (0.21)					0.23	0.16	0.40	1.76[5]				
1.2	—4.15 (4.90)	1.05 (0.30)	1.48 (0.55)				0.57	0.48	0.91	3.46[3]	2.67[3]			
						For 1963-66 (with 1960-63 as base)								
1.1	−0.84 (1.90)				0.07 (0.04)		0.24	0.17	0.44				1.82[5]	

TABLE 5.3 **Area Responses of Rice**

Stepwise Regression Results
Dependent Variable : Growth Rate in Area
No. of Independent Variables : 5
Relevant Number of Samples for Regression : 15
North Eastern Peninsular Plateau Region

Equation Number	Cons-tant	Regression Co-efficient of					R^2	$\bar{R}^2$	Constant	T Value				
		X_1	X_2	X_3	X_4	X_5				X^1	X_2	X_3	X_4	X_5
1	2	3	4	5	6	7	8	9	10	11	12	13	14	15
For 1970-72 (with 1960-63 average as base)														
1.1	—8.6 (6.20)					0.93 (0.93)	0.37	0.32	1.39					2.77[3]
For 1970-72 (with 1963-66 average as base)														
1.1	8.03 (4.38)					—0.29 (0.19)	0.15	0.08	1.83[5]					1.53
1.2	11.42 (0.32)			—0.50 (0.32)		—0.37 (0.18)	0.29	0.17	2.52			1.53[5]		1.98[4]
1.3	11.06 (4.69)	0.20 (0.25)		—0.48 (0.33)		—0.41 (0.19)	0.33	0.15	2.35	0.81*		1.43[5]		2.09[4]

TABLE 5.3 (*Contd.*)

1	2	3	4	5	6	7	8	9	10	11	12	13	14	15
For 1963-66 (with 1960-63 average as base)														
1.1	—3.04					—0.24	0.06	—0.01	1.05					0.92*
	(2.88)					(0.26)								
1.2	—6.43			1.77		—0.52	0.24	0.11	2.10		1.70[5]			1.76[5]
	(3.05)			(1.04)		(0.29)								
1.3	—12.6			2.00	0.16	—0.76	0.33	0.15	2.53		1.93[5]		1.23*	2.18[4]
	(4.97)			(1.03)	(0.13)	(0.35)								
1.4	—12.04		0.21		0.15	—0.78	0.37	0.12	2.33		0.78*		1.12*	2.10[4]
	(5.16)		(0.27)		(0.13)	(0.35)								

TABLE 5.4 Area Responses of Rice

Stepwise Regression Results
Dependent Variable : Growth Rate in Area
No. of Independent Variables ; 5
Relevant Number of Samples for Regression : 10
Assam Plains Region

Equation Number	Constant	Regression Co-efficient of					R^2	$\bar{R}^2$	Constant	T Value				
		X_1	X_2	X_3	X_4	X_5				X_1	X_2	X_3	X_4	X_5
1	2	3	4	5	6	7	8	9	10	11	12	13	14	15
For 1970-72 (with 1960-63 average as base)														
1.1	—11.03 (7.48)			8.30 (2.32)			0.64	0.59	1.47			3.58[4]		
For 1970-72 (with 1963-66 average as base)														
1.1	3.27 (2.50)		—0.98 (0.40)				0.46	0.38	1.30		2.44[5]			
For 1963-66 (with 1960-63 average as base)														
1.1	14.77 (3.19)		6.76 (1.80)				0.66	0.62	4.62			3.74[4]		
1.2	23.42	—3.20	14.29				0.81	0.75	3.95		2.15[5]	3.78[4]		

TABLE 5.5 Area Responses of Rice

Stepwise Regression Results
Dependent Variable : Growth Rate in Area
No. of Independent Variables : 5
Relevant Number of Samples for Regression : 10
Ganges Delta Region and Assam Plains Region

Equation Number	Constant	Regression Co-efficient of X_1	X_2	X_3	X_4	X_5	R^2	$\bar{R}^2$	Constant	T Value X_1	X_2	X_3	X_4	X_5
1	2	3	4	5	6	7	8	9	10	11	12	13	14	15
For 1970-72 (with 1963-66 average as base)														
1.1	3.62 (2.25)					0.15 (0.10)	0.10	0.05	1.60					1.38[5]
1.2	3.71 (2.36)	0.07 (0.09)				0.17 (0.11)	0.14	0.02	1.56	0.77*				1.53[5]
For 1963-66 (with 1960-63 average as base)														
1.1	2.79 (1.58)					0.16 (0.14)	0.07	0.01	1.75					1.13*
1.2	2.63 (1.57)	0.12 (0.10)				—0.17 (0.14)	0.15	0.03	1.67					1.17*
1.3	—7.77 (1.10)	0.12 (0.10)			0.31 (0.85)	0.57 (1.10)	0.15	—0.02	0.23	1.12*			0.36*	0.51*
1.4	—570.90 (377.09)	0.06 (0.10)		—27.34 (16.18)	16.91 (9.85)	—22.50 (13.15)	0.31	0.09	1.51	0.59		1.68[5]	1.71[5]	1.72[5]

For 1970-72 (with 1960-63 average as base)

1.1	9.39	0.18				0.04	0.22	2.78	0.90*			
	(3.36)	(0.20)										
1.2	—12.45	0.67	5.78			0.27	0.17	1.04	2.29[4]	2.13[4]		
	(2.70)	(0.29)	(2.70)									
1.3	—18.45	0.65	4.06	0.08		0.28	0.13	0.58	2.12[4]	0.94*	0.52*	
	(31.05)	(0.30)	(4.31)	(0.16)								
1.4	—146.43	0.68	—0.15	1.78	—3.75	0.35	0.15	0.61	2.24[3]	0.89*	1.22	0.61*
	(239.64)	(0.30)	(16.80)	(1.54)	(3.19)							

Note : 1. As the number of samples for regression were not sufficient to run stepwise regression Assam Plains region is combined with Ganges Delta. The results have been provided separately also for Assam Plains. See Table 5.4.

TABLE 5.6 **Area Responses of Rice**

Stepwise Regression Results
Dependent Variable : Growth in Area
No. of Independent Variables : 5
Relevant Number of Samples for Regression : 20
Lower and Upper Ganges Plain Region

Equation Number	Constant	Regression Co-efficient of					R^2	$\bar{R}^2$	Constant	TV alue				
		X_1	X_2	X_3	X_4	X_5				X_1	X_2	X_3	X_4	X_5
1	2	3	4	5	6	7	8	9	10	11	12	13	14	15
For 1970-72 (with 1963-66 average as base)														
1.1	7.46 (3.49)		—0.42 (0.20)				0.19	0.14	2.13		2.07[4]			
For 1970-72 (with 1960-63 average as base)														
1.1	4.79 (0.18)				0.06 (0.18)		0.38	0.35					3.38[1]	

For 1963-66 (with 1960-63 average as base)

1.1	8.39				—0.32	0.09	0.03	3.28				1.31*
	(2.55)				(0.24)							
1.2	11.97			—0.08	—0.32	0.16	0.05	3.00			1.16*	1.30*
	(3.97)			(0.07)	(0.24)							
1.3	12.23	—0.08		—0.09	—0.29	0.22	0.07	3.09	1.10*		1.15*	1.21*
	(3.96)	(0.08)		(0.08)	(0.24)							
1.4	11.85	—0.08	0.61	—0.08	—0.29	0.24	0.02	2.90	1.08*	0.57*	1.06*	1.16*
	(4.07)	(0.08)	((1.06)	(0.07)	(0.25)							

TABLE 5.7 **Area Responses of Rice**

Stepwise Regression Results
Dependent Variable : Growth in Area
No. of Independent Variables ; 5
Relevant Number of Samples for Regression : 74
All Rice Regions

Equation Number	Constant	Regression Co-efficient of					R^2	$\bar{R}^2$	Constant	T Value				
		X_1	X_2	X_3	X_4	X_5				X_1	X_2	X_3	X_4	X_5
1	2	3	4	5	6	7	8	9	10	11	12	13	14	15
For 1970-72 (with 1960-63 average as base)														
1.1	—1.90				0.07		0.16	0.15	0.60				3.74[1]	
	(3.18)				(0.01)									
For 1970-72 (with 1963-66 average as base)														
1.1	5.33		0.53				0.07	0.27	3.42			2.45[2]		
	(1.55)													

For 1963-66 (with 1960-63 average as base)

1.1	3.20			—0.13	0.02	0.01	3.06			1.29[5]
	(1.04)			(0.10)						
1.2	3.19		0.32	—0.12	0.03	0.004	3.06		0.81*	1.19*
	(1.04)		(0.39)	(0.10)						
1.3	3.27	—0.03	0.34	—0.12	0.05	0.004	3.09	0.62*	0.87*	1.16*
	(1.05)	(0.05)	(0.59)	(0.10)						

TABLE 5.8 Yield Responses of Rice

Stepwise Regression Results
Dependent Variable : Yield Per Hectare
No. of Independent Variables : 6
Relevant No. of Samples for Regression : 10
West Coast Region

Equation Number	Constant	Regression Co-efficient of X_1	X_2	X_3	X_4	X_5	X_6	R^2	$\bar{R}^2$	Constant	T Value X_1	X_2	X_3	X_4	X_5	X_6
1	2	3	4	5	6	7	8	9	10	11	12	13	14	15	16	17
								For 1967-68								
1.1	1254.46 (213.78)		448.78 (213.78)					0.35	0.27	14.91		2.09[4]				
								For 1970-71								
1.1	1008.45 (96.60)		449.15 (86.23)					0.77	0.74	10.43		5.20[3]				
1.2	1204.29 (164.88)		615.17 (85.92)			—17.07 (6.39)		0.88	0.85	7.30		6.84[2]			2.68[5]	

1.3	1364.14 (5.87)	—6.93 (4.30)	658.99 (11.59)		—18.77 (5.87)	0.92	0.88	7.38	1.61*	7.69[2]		3.19[4]

Note : Figures in Parenthesis denote the standard error

1. Significant at 99% level
2. Significant at 98% level
3. Significant at 95% level
4. Significant at 90% level
5. Significant at 80% level

*Insignificant

List of Independent Variables

X_1 Fertilizer consumption in kgs per hectare

X_2 No. of tractors per thousand/hectare of total cropped area

X_3 No. of pumpsets and engines per thousand hectare

X_4 Percentage of area under high yielding variety of the crop to area under the crop

X_5 Percentage of wheat area irrigated to total area under wheat

X_6 Price of crop (per kilogram)

TABLE 5.9 Yield Responses of Rice

Stepwise Regression Results
Dependent Variable : Yield Per Hectare
No. of Independent Variables : 6
Relevant No. of Samples for Regression : 16
East Coast Region

Equation Number	Constant	Regression Co-efficient of						R^2	$\bar{R}^2$	Constant	T Value					
		X_1	X_2	X_3	X_4	X_5	X_6				X_1	X_2	X_3	X_4	X_5	X_6
1	2	3	4	5	6	7	8	9	10	11	12	13	14	15	16	17
								For 1967-68								
1.1	891.21 (78.49)	12.56 (2.53)						0.63	0.61	11.35	4.94[3]					
								For 1970-71								
1.1	176.40 (127.78)	20.82 (3.49)						0.71	0.69	6.07	5.95[3]					
1.2	729.11 (151.17)	16.74 (3.65)			5.91 (2.75)			0.79	0.75	4.82	4.58[3]					2.14[4]

TABLE 5.10 Yield Responses of Rice

Stepwise Regression Results
Dependent Variable : Yield Per Hectare
No. of Independent Variables : 6
Relevant No. of Samples for Regression : 23
North Eastern Peninsular Plateau Region

Equation Number	Constant	Regression Co-efficient of						R^2	$\bar{R}^2$	Constant	T Value					
		X_1	X_2	X_3	X_4	X_5	X_6				X_1	X_2	X_3	X_4	X_5	X_6
1	2	3	4	5	6	7	8	9	10	11	12	13	14	15	16	17
								For 1967-68								
1.1	622.85				5.74			0.19	0.15	11.52			20.25[1]			
	(54.04)				(2.54)											
								For 1970-71								
1.1	692.85				7.58			0.52	0.50	17.60				4.82[1]		
	(39.36)				(1.57)											
1.2	625.44				7.15		120.86	0.63	0.59	14.04				5.45[1]		2.38[2]
	(50.60)				(1.42)		(50.60)									

TABLE 5.11 **Yield Responses of Rice**

Stepwise Regression Results
Dependent Variable : Yield Per Hectare
No. of Independent Variables : 6
No. of Samples for Regression : 18
Assam Plains Region and Ganges Delta

Equation Number	Constant	Regression Co-efficient of						R^2	$\bar{R}^2$	Constant	T Value					
		X_1	X_2	X_3	X_4	X_5	X_6				X_1	X_2	X_3	X_4	X_5	X_6
1	2	3	4	5	6	7	8	9	10	11	12	13	14	15	16	17
								For 1967-68								
1.1	765.23 (117.52)	38.00 (20.28)						0.17	0.12	6.54	1.87[4]					
								For 1970-71								
1.1	1035.86 (53.78)	15.78 (3.29)						0.58	0.56	19.26	4.79[1]					

Table 5.12 Yield Responses of Rice

Stepwise Regression Results
Dependent Variable : Yield Per Hectare
No. of Independent Variables : 6
Relevant Number of Samples for Regression : 20
Lower and Upper Ganges Plain Region

Equation Number	Cons-tant	Regression Co-efficient of						R^2	$\bar{R}^2$	Constant	T Value					
		X_1	X_2	X_3	X_4	X_5	X_6				X_1	X_2	X_3	X_4	X_5	X_6
1	2	3	4	5	6	7	8	9	10	11	12	13	14	15	16	17
								For 1967-68								
1.1	541.27 (41.60)		188.25 (42.20)					0.52	0.49	13.01		4.45[1]				
1.2	1033.95 (160.43)		199.06 (34.59)			—344.90 (108.80)		0.70	0.66	6.44		5.75[1]			3.17[1]	
1 3	951.76 (147.84)				2.02 (0.90)	—310.60 (99.14)		0.77	0.72	6.43		4.54[1]		2.23[3]	3.13[1]	

TABLE 5.12 (*Contd.*)

1	2	3	4	5	6	7	8	9	10	11	12	13	14	15	16	17
								For 1970-71								
1.1	1266.93	—23.28						0.11	0.06	4.96		1.54[5]				
	(255.17)	(15.07)														
1.2	1353.79	—16.89	—238.51					0.16	0.06	4.07		1.02*	0.94*			
	(331.97)	(16.55)	(251.46)													
1.3	1694.27	—0.76	331.45			—0.48		0.20	0.06	2.57		0.03*	1.23*		0.97*	

TABLE 5.13 **Yield Responses of Rice**

Stepwise Regression Results
Dependent Variable : Yield Per Hectare
No. of Independent Variables : 6
Relevant Number of Samples for Regression : 87
All Rice Regions

Equation Number	Constant	Regression Co-efficient of						R^2	$\bar{R}^2$	Constant	T Value					
		X_1	X_2	X_3	X_4	X_5	X_6				X_1	X_2	X_3	X_4	X_5	X_6
1	2	3	4	5	6	7	8	9	10	11	12	13	14	15	16	17
For 1967-68																
1.1	685.62				6.46			0.28	0.55	13.77				5.55[1]		
	(49.78)				(1.10)											
1.2	651.92	8.69			4.56			0.37	0.35	11.15	3.32			3.84[1]		
	(1.18)	(2.61)			(1.18)											

TABLE 5.13 (*Contd.*)

1	2	3	4	5	6	7	8	9	10	11	12	13	14	15	16	17
								For 1970-71								
1.1	873.17	15.87						0.26	0.25	14.18	5.52[1]					
	(61.55)	(2.87)														
1.2	775.56	10.51			4.99			0.34	0.32	9.17	3.28[1]			3.22[1]		
	(84.51)	(3.19)			(1.54)											
1.3	611.88	9.96			5.06		252.45	0.38	0.36	5.55	3.18[1]			3.85[1]		2.32[3]
	(110.10)	(3.12)			(1.50)		(108.68)									
1.4	601.17	7.22			4.44	5.77	243.19	0.40	0.37	4.87	2.05[3]			2.88[1]	1.62[5]	2.25[3]
	(123.32)	(3.52)			(1.54)	(3.54)	(107.76)									

TABLE 5.14 Area Responses of Rice

Stepwise Regression Results
Dependent Variable : Growth in Area
No. of Independent Variables : 5
Relevant Number of Samples for Regression : 36
Upper Ganges Plains Region

Equation Number	Constant	Regression Co-efficient of					R^2	$\bar{R}^2$	Constant	T Value				
		X_1	X_2	X_3	X_4	X_5				X_1	X_2	X_3	X_4	X_5
1	2	3	4	5	6	7	8	9	10	11	12	13	14	15
For 1970-72 (1960-63 average as base)														
1.1	44.45 (5.17)	0.71 (0.37)					0.09	0.07	8.59	1.91[4]				
For 1970-72 (1963-66 average as base)														
1.1	39.49 (5.33)		1.34 (0.48)				0.18	0.15	7.40	2.76[1]				
For 1963-66 (1960-63 average as base)														
1.1	0.78 (1.21)		—0.35 (0.23)				0.06	0.03	0.64	1.49[5]				

TABLE 5.14 (*Contd.*)

1	2	3	4	5	6	7	8	9	10	11	12	13	14	15
1.2	—2.91		—0.33		0.66		0.08	0.02	0.69	1.41[5]			0.91[4]	
	(4.18)		(0.23)		(0.07)									

Note : Figures in Parenthesis denote the standard error

1. Significant at 99% level
2. Significant at 98% level
3. Significant at 95% level
4. Significant at 90% level
5. Significant at 80% level

* Insignificant

Equation Numbers 1.1, 1.2 etc., indicate steps of regression

List of Independent Variables

X_1 Growth rate in gross cropped area

X_2 Variation in gross area irrigated as a percentage of gross cropped area

X_3 Variation in percentage of area irrigated of the crop to total area under crop

X_4 Growth rate in yield (Rupees per hectare)

X_5 Growth rate in yield (kgs. per hectare)

TABLE 5.15 Area Responses of Wheat

Stepwise Regression Results
Dependent Variable : Growth in Area
No. of Independent Variables : 5
Relevant Number of Samples for Regression : 12
Lower Ganges Plain Region

Equation Number	Constant	Regresssion Co-efficient of					R^2	$\bar{R}^2$	Constant	T Value				
		X_1	X_2	X_3	X_4	X_5				X_1	X_2	X_3	X_4	X_5
1	2	3	4	5	6	7	8	9	10	11	12	13	14	15
For 1970-72 (1960-63 average as base)														
1.1	0.89		2.99				0.58	0.54	7.71		3.71[1]			
	(11.58)		(0.80)											
For 1970-72 (1963-66 average as base)														
1.1	17.25			4.18			0.62	0.58	0.57			4.08[1]		
	(29.77)			(1.02)										
For 1963-66 (1960-63 average as base)														
1.1	8.80				—0.21		0.17	0.08	1.07				1.43*	
	(8.16)				(0.15)									

TABLE 5.15 (*Contd.*)

1	2	3	4	5	6	7	8	9	10	11	12	13	14	15
1.2	5.13			0.81	—0.18		0.23	0.06	0.58			0.87*	1.17*	
	(8.78)			(0.93)	(0.15)									
1.3	6.17		—0.89	0.94	—0.19		0.26	0.01	0.66		0.55*	0.94*	1.20*	
	(9.34)		(1.64)	(0.99)	(0.16)									
1.4	22.28		—2.27	1.12	—0.39	0.79	0.34	—0.02	1.45		1.02*	1.47*	0.93*	
	(15.29)		(2.22)	(1.02)	(0.26)									
1.5	49.31	2.55	—4.00	1.36	—0.73	2.09	0.54	0.15	2.69	1.60	1.77[5]	1.45	2.28	1.87[5]

TABLE 5.16 **Area Responses of Wheat**

Stepwise Regression Results
Dependent Variable : Growth in Area
No. of Independent Variations : 5
Relevant Number of Samples for Regression : 15
Punjab Plains Region

Equation Number	Constant	Regression Co-efficient of X_1	X_2	X_3	X_4	X_5	R^2	$\bar{R}^2$	Constant	T Value X_1	X2	X_3	X_4	X_5
1	2	3	4	5	6	7	8	9	10	11	12	13	14	15
					For 1970-72 (1960-63 average as base)									
1.1	35.75		1.30				0.19	0.13	2.74		1.80[5]			
	(0.72)		(0.72)											
					For 1970-72 (1963-66 average as base)									
1.1	54.59					—0.18	0.15	0.06	6.52					1.42[5]
	(8.37)					(0.13)								
1.2	41.01		0.91			—0.16	0.24	0.11	3.22		1.30*			1.13*
	(12.71)		(0.70)			(0.12)								

TABLE 5.16 (*Contd.*)

1	2	3	4	5	6	7	8	9	10	11	12	13	14	15
1.3	61.16		1.40	—1.06		—0.29	0.36	0.18	3.04		1.85[5]	1.42[5]		1.93[5]
	(20.07)		(0.75)	(0.74)		(0.15)								
1.4	56.51		1.37	—1.27	0.07	—0.29	0.38	0.13	2.05		1.75[5]	1.49[5]	0.56*	1.89[5]
	(27.51)		(0.78)	(0.85)		(0.15)								
1.5	56.83	—0.29	1.28	—1.23	0.05	—0.26	0.39	0.06	1.83	0.47*	1.53[5]	1.38*	0.67*	1.58[5]
For 1963-66 (1960-63 average as base)														
1.1	2.76					(0.41)	0.39	0.34	0.97					2.89[3]
	(2.83)					(0.41)								

TABLE 5.17 Area Responses of Wheat

Stepwise Regression Results
Dependent Variable : Growth in Area
No. of Independent Variables : 5
Relevant Number of Samples for Regression ; 16
North Eastern Peninsular Plateau of Region

Equation Number	Constant	Regression Co-efficient of					R^2	$\bar{R}^2$	Constant	T Value				
		X_1	X_2	X_3	X_4	X_5				X_1	X_2	X_3	X_4	X_5
1	2	3	4	5	6	7	8	9	10	11	12	13	14	15
For 1970-72 (1960-63 average as base)														
1.1	2.61			1.02			0.38	0.34	0.33			2.97[2]		
	(7.88)			(0.34)										
1.2	4.30		—4.24	1.81			0.57	0.50	0.39		2.38[3]	4.07[1]		
	(10.88)		(1.73)	(0.44)										
For 1970-72 (1963-66 average as base)														
1.1	22.86	0.69					0.20	0.14	3.18	1.87[4]				
	(7.17)	(0.36)												

TABLE 5.17 (*Contd.*)

1	2	3	4	5	6	7	8	9	10	11	12	13	14	15
For 1963-66 (1960-63 average as base)														
1.1	—8.11				0.12		0.04	0.02	1.81				0.80*	
	(4.46)				(0.16)									
1.2	—7.78			—0.18	0.16		0.07	—0.06	1.58			0.67*	0.95*	
	(4.92)			(0.28)	(0.17)									
1.3	—10.18		2.26	—0.36	0.23		0.16	—0.04	1.87		1.09*	1.13*	1.29*	

TABLE 5.18 **Area Responses of Wheat**

Stepwise Regression Results
Dependent Variable : Growth in Area
No. of Independent Variables : 5
Relevant Number of Samples for Regression : 22
North Peninsular Plateau Region

Equation Number	Constant	Regression Co-efficient of					R^2	$\bar{R}^3$	Constant	T Value				
		X_1	X_2	X_3	X_4	X_5				X_1	X_2	X_3	X_4	X_5
1	2	3	4	5	6	7	8	9	10	11	12	13	14	15
For 1970-72 (1960-63 average as base)														
1.1	−1.41	0.99					0.75	0.74	0.52	7.83[1]				
	(2.70)	(0.12)												
For 1970-72 (1963-66 average as base)														
1.1	2.71	0.64					0.41	0.39	0.89	3.97[1]				
	(3.03)	(0.64)												
For 1963-66 (1960-63 average as base)														
1.1	−5.75	4.52					0.34	0.30	1.60	3.06[1]				
	(3.58)	(1.47)												
1.2	−4.44	4.87		−1.14			0.34	0.41	1.29	3.56[1]		2.07[4]		

TABLE 5.19 **Area Responses of Wheat**

Stepwise Regression Results
Dependent Variable : Growth in Area
No. of Independent Variables : 5
Relevant Number of Samples for Regression : 9
North Western Transitional Region

Equation Number	Constant	Regression Co-efficient of					R^2	$\bar{R}^2$	Constant	T Value				
		X_1	X_2	X_3	X_4	X_5				X_1	X_2	X_3	X_4	X_5
1	2	3	4	5	6	7	8	9	10	11	12	13	14	15
For 1970-72 (1960-63 average as base)														
1.1	20.25		1.66				0.59	0.53	4.43		2.98*			
	(4.57)		(0.55)											
1.2	8.13		1.77			0.28	0.72	0.62	0.96		3.50[5]	1.55*		
	(8.64)		(0.50)			(0.18)								
1.3	3.38	0.49	1.85			0.20	0.76	0.61	0.27	0.98*	3.60[5]			1.02*
	(12.24)	(0.50)	(0.51)			(0.20)								

1.4	3.00	0.53	1.51	0.32	0.64	0.79	0.52	0.18	0.94*	1.65*	0.47*	0.53*
	(16.51)	(0.56)	(0.91)	(0.68)								
For 1970-72 (1963-66 average as base)												
1.1	19.89	1.24				0.40	0.30	1.84	2.00			
	(10.75)	(0.62)										
1.2	6.11	2.83	4.86			0.96	0.95	1.73	12.55[1]	9.63[4]		
	(5.52)	(0.22)	(0.50)									
1.3	−4.81	2.82	5.02	−0.37		0.98	0.97	1.79	16.81[3]	13.15[3]	2.24*	
	(2.67)	(0.16)	(0.38)	(0.16)								
1.4	−5.79	2.79	4.99	−0.44	2.79	0.98	0.97	1.20	13.94[8]	11.58[4]	1.84*	0.48*
For 1963-66 (1960-63 average as base)												
1.1	−13.54	2.60				0.88	0.86	8.77	6.68[4]			
	(1.54)	(0.38)										
1.2	−12.70	2.68		−0.18		0.90	0.86	7.26	6.96[4]	1.15*		
	(1.74)	(0.28)		(0.16)								

TABLE 5.19 (*Contd.*)

1	2	3	4	5	6	7	8	9	10	11	12	13	14	15
1.3	— 9.01	2.44		0.18	—0.06		0.92	0.87	1.89	5.26[5]	0.77*	0.30	1.27*	
	(3.41)	(0.44)		(0.15)	(0.06)									
1.4	— 7.82	2.45	—0.28	—0.06	—0.09		0.93	0.85	1.89	5.26[5]	0.77*	0.30	1.27*	
	(4.13)	(0.46)	(0.36)	(0.22)	(0.77)									
1.5	— 7.66	2.44	—0.36	—0.10			0.93	0.89	2.33	5.98[5]	1.48*	1.67*		
	(3.28)	(0.40)	(0.24)	(0.06)										
1.6	9.25	2.19	—0.30		—0.46		0.94	0.81	0.41	1.23*	0.83*	0.35*	0.55*	
	(29.53)	(0.64)	(0.28)		(0.64)									

TABLE 5.20 Area Responses of Wheat

Stepwise Regression Results
Dependent Variable : Growth in Area
No. of Independent Variables : 5
Relevant Number of Samples for Regression : 109
All Wheat Regions

Equation Number	Constant	Regression Co-efficient of					R^2	$\bar{R}^2$	Constant	T Value				
		X_1	X_2	X_3	X_4	X_5				X_1	X_2	X_3	X_4	X_5
1	2	3	4	5	6	7	8	9	10	11	12	13	14	15
For 1970-72 (1960-63 average as base)														
1.1	21.48			1.13										
	(5.48)			(0.21)			0.20	0.19	3.91			5.20[1]		
1.2	16.28			0.98		0.21	0.24	0.23	2.54			4.47[1]		2.46[2]
	(6.38)			(0.22)		(0.08)								
1.3	14.98		0.74	0.76		0.21	0.26	0.24	1.95		1.67[4]	2.97[1]		2.51[2]
	(7.65)		(0.44)	(0.25)		(0.06)								
1.4	9.52	0.49	0.93	0.71		0.18	0.28	0.26	1.15	1.88[4]	2.68[1]	2.79[1]		1.19*
	(8.26)	(0.26)	(0.44)	(0.25)		(0.08)								

TABLE 5.20 (*Contd.*)

1	2	3	4	5	6	7	8	9	10	11	12	13	14	15
						For 1970-72 (1960-63 average as base)								
1.1	19.05			1.81			0.23	0.22	3.23			5.78[1]		
	(5.88)			(0.31)										
1.2	26.80			1.95	—0.08		0.25	0.24	3.30			6.09[1]	1.76[4]	
	(8.09)			(0.32)	(0.05)									
1.3	23.46			1.84	—0.12	0.18	0.27	0.25	2.42			5.70[1]	2.32[3]	1.80*
	(9.67)			(0.32)	(0.05)	(0.10)								
1.4	18.53	0.65		1.87	—0.15	0.21	0.30	0.27	1.79	1.96		5.86[1]	2.81[1]	2.07[3]
	(10.28)	(0.33)		(0.31)	(0.31)	(0.10)								
1.5	16.33	0.78	0.92	1.59	—0.17	0.22	0.32	0.29	1.46	2.31	1.72	4.51	3.12[1]	2.25[3]
	(11.13)	(0.33)	(0.52)	(0.35)	(0.05)	(0.10)								
						[*For 1963-66 (1960-63 average as base)*								
1.1	0.55					0.18	0.04	0.03	0.46					2.26[3]
	(1.19)					(0.08)								
1.2	4.31				—0.08	0.26	0.06	0.07	2.04				1.89[4]	2.92[1]
	(2.11)				(0.04)	(0.09)								

TABLE 5.21 **Yield Responses of Wheat**

Stepwise Regression Results
Dependent Variable : Yield Per Hectare
No. of Independent Variables : 6
Relevant Number of Samples for Regression : 36
Upper Ganges Plain Region

Equation Number	Constant	Regression Co-efficient of						R^2	$\bar{R}^2$	Constant	T Value					
		X_1	X_2	X_3	X_4	X_5	X_6				X_1	X_2	X_3	X_4	X_5	X_6
								For 1967-68								
1.1	762.29 (127.00)				7.02 (2.04)			0.25	0.23	6.00				3.44[1]		
								For 1970-71								
1.1	944.36 (151.73)				5.14 (7.02)			0.16	0.13	6.22				2.54[2]		

Note : Figures in Parenthesis denote the standard error]

1. Significant at 99% level
2. Significant at 98% level
3. Significant at 9. % level
4. Significant at 90 , level
5. Significant at 80% level

*Insignificant

List of Independent Variables

X_1 Fertiliser Consumption in kilograms per hectare
X_2 No. of tractors per thousand/hectares of total cropped area
X_3 No. of pumpsets and engines per thousand hectares
X_4 Percentage of area under high yielding variety of the crop to area under the crop
X_5 Percentage of wheat area irrigated to total area under wheat
X_6 Price of crop (per kilogram)

TABLE 5.22 Yield Responses of Wheat

Stepwise Regression Results
Dependent Variable : Yield Per Hectare
Number of Independent Variables : 6
Relevant Number of Samples for Regression : 17
Lower Ganges Plain Region

Equation Number	Constant	Regression Co-efficient of						R^2	$\bar{R}^2$	Constant	T Value					
		X_1	X_2	X_3	X_4	X_5	X_6				X_1	X_2	X_3	X_4	X_5	X_6
								For 1967-68								
1.1	715.35			63.03				0.30	0.23	15.59				2.07[4]		
	(45.87)			(30.30)												
1.2	737.09			112.50		—2.50		0.54	0.44	12.35				3.28[1]	2.10[4]	
	(59.66)			(34.23)		(1.55)										
1.3	719.01		65.39	97.33		—2.32		0.65	0.53	12.21		1.62[5]		2.96[1]	2.17[4]	
	(58.88)		(40.03)	(32.80)		(1.06)										
								For 1970-71								
1.1	1957.70		—665.58					0.42	0.36	8.34		2.71[3]				
	(234.66)		(244.91)													

TABLE 5.23 Yield Responses of Wheat

Stepwise Regression Results
Dependent Variable : Yield Per Hectare
Number of Independent Variables : 6
Relevant Number of Samples for Regression : 15
Punjab Plains Region

Equation Number	Constant	Regression Co-efficient of						R^2	$\bar{R}^2$	Constant	T Value					
		X_1	X_2	X_3	X_4	X_5	X_6				X_1	X_2	X_3	X_4	X_5	X_6
1	2	3	4	5	6	7	8	9	10	11	12	13	14	15	16	17
								For 1967-68								
1.1	1229.88 (143.82)			49.36 (15.16)				0.44	0.40	8.55			5.25[2]			
								For 1970-71								
1.1	660.13 (324.26)					23.02		0.62	0.59	2.03					4.63[1]	

TABLE 5.24 **Yield Responses of Wheat**

Stepwise Regression Results
Dependent Variable : Yield Per Hectare
Number of Independent Variables : 6
Relevant Number of Samples for Regression : 16
North Eastern Peninsular Plateau Region

Equation Number	Constant	Regression Co-efficient of						R^2	$\bar{R}^2$	Constant	T Value					
		X_1	X_2	X_3	X_4	X_5	X_5				X_1	X_2	X_3	X_4	X_5	X_6
1	2	3	4	5	6	7	8	9	10	11	12	13	14	15	16	17
								For 1967-68								
1.1	614.49				7.08			0.53	0.49	11.33				3.98[1]		
	(54.19)				(1.77)											
								For 1970-71								
1.1	754.08				5.73			0.50	0.47	12.39				3.81[1]		
	(60.82)				(1.50)											

TABLE 5.25 Yield Responses of Wheat

Stepwise Regression Results
Dependent Variable : Yield Per Hectare
Number of Independent Variables : 6
Relevant Number of Samples for Regression : 22
North Peninsular Plateau Region

Equation Number	Constant	Regression Co-efficient of						R^2	$\bar{R}^2$	Constant	T Value					
		X_1	X_2	X_3	X_4	X_5	X_6				X_1	X_2	X_3	X_4	X_5	X_6
								For 1967-68								
1.1	1243.34 (499.24)				—37.24 (45.98)			0.03	—0.01	2.49				0.50*		
1.2	1648.9 (648.6)			—151.6 (172.7)	—43.7 (46.8)			0.06	—0.02	2.54			0.89*	0.93*		
1.3	1467.1 (724.6)		670.2 (1116.5)	—147.3 (175.8)	—44.8 (47.68)			0.08	—0.06	2.02		0.60*	0.83*	0.94*		
1.4	1396.2 (901.7)		789.2 (1178.0)	—151.0 (680.2)	61.4 (63.1)	84.1 (202.0)		0.09	—0.11	1.54		0.66*	0.83*	0.97*		0.41
								For 1970-71								
1.1	676.84 (73.90)				7.00 (3.14)			0.35	0.29	9.15				2.24[3]		

TABLE 5.26 **Yield Responses of Wheat**

Stepwise Regression Results
Dependent Variable : Yield Per Hectare
Number of Independent Variables : 6
Relevant Number of Samples for Regression : 9
North Western Transitional Region

Equation Number	Constant	Regression Co-efficient of						R^2	$\bar{R}^2$	Cons-tant	T Value					
		X_1	X_2	X_3	X_4	X_5	X_6				X_1	X_2	X_3	X_4	X_5	X_6
1	2	3	4	5	6	7	8	9	10	11	12	13	14	15	16	17
								For 1967-68								
1.1	682.313 (93.82)			160.87 (60.13)				0.54	0.46	7.27			2.42[5]			
1.2	1860.42 (467.45)			155.34 (43.67)			—1401.39	0.80	0.72	3.97			3.55[4]			2.53[5]
								For 1970-71								
1.1	889.10 (208.77)				5.88 (2.83)			0.41	0.32	4.25				1.07*		
1.2	830.22 (176.83)		—393.11 (147.37)		0.90 (2.29)			0.76	0.66	4.69		2.67[5]		3.89[4]		

TABLE 5.27 **Yield Responses of Wheat**

Stepwise Regression Results
Dependent Variable : Yield Per Hectare
No. of Independent Variables : 6
Relevant Number of Samples for Regression : 6 111
All Wheat Regions

Equation Number	Constant	Regression Co-efficient of						R^2	$\bar{R}^2$	Constant	T Value					
		X_1	X_2	X_3	X_4	X_5	X_6				X_1	X_2	X_3	X_4	X_5	X_6
1	2	3	4	5	6	7	8	9	10	11	12	13	14	15	16	17
								For 1967-68								
1.1	873.07		334.53					0.06	0.05	8.78		2.80[3]				
	(99.40)		(119.21)													
1.2	722.76		268.17		4.59			0.09	0.07	5.10		2.17[3]		1.84[4]		
	(141.66)		(123.29)		(2.48)											
1.3	1567.06		266.11		3.46		—948.19	0.33	0.08	2.41		2.16[3]		1.31[5]		1.26*
	(648.86)		(122.90)		(2.48)											

TABLE 5.27 (*Contd.*)

Equation Number	Cons-tant	Regression Co-efficient of						R^2	$\bar{R}^2$	Cons-tant	T Value					
		X_1	X_2	X_3	X_4	X_5	X_6				X_1	X_2	X_3	X_4	X_5	X_6
1	2	3	4	5	6	7	8	9	10	11	12	13	14	15	16	17
1.4	1660.65		211.99	17.94	3.16		1067.30	0.33	0.08	2.49		1.49*	0.77	1.18*		1.56[5]
	(66.85)		(141.68)	(23.20)	(2.67)		(760.48)									
1.1	701.34					17.43		0.55	0.54	12.00					11.55[1]	
	(50.40)					(1.50)										
1.2	718.20		13.17			64.32		0.63	0.63	11.33		5.11[1]			8.27[1]	
	(63.30)		(1.59)			(12.57)										
1.3	599.88		54.63		4.64			0.67	0.66	5.81		4.46[1]		3.61[1]	6.37[1]	
	(88.03)		(12.22)		(1.11)											
1.4	581.91		33.58	5.00	4.35	9.22		0.69	0.68	5.90		2.16[3]	2.20[3]	3.93[1]	5.25[1]	
	(98.56)		(15.31)	(2.26)	(1.10)	(1.75)										
1.5	579.62	—3.07	38.17	5.78	4.70	9.36		0.65	0.67	4.90	0.07	2.31[3]	2.32[3]	3.92[1]	5.29[1]	
	(118.08)	(4.02)	(16.47)	(2.49)	(2.20)	(1.76)										

Note : Figures in paranthesis denote the standard error

1. Significant at 99% level
2. Significant at 98% level
3. Significant at 95% level
4. Significant at 90% level
5. Significant at 80% level

List of Independent Variables

X_1 Fertilizer consumption in kilogram

X_2 No. of tractors per thousand/ha of total cropped area

X_3 No. of pumpsets and engines per thousand hectares

X_4 Percentage of area under high yielding variety of the crop to area under the crop

X_5 Percentage of wheat area irrigated to total area under wheat

X_6 Price of crop (per Kilogram)

6

Regional Patterns and Variations in Area and Productivity of Rice and Wheat—Case Studies

6.1 INTRODUCTION

This chapter is based on a cross sectional study of eleven villages of rice and wheat for the reference period 1975-76. These samples were selected on the basis of proportional sampling.[1] This is a method of sample selection where a unit in the universe is selected according to its relative population with respect to total population of all the units in the universe. In other words, the probability of selecting a unit will vary directly with the population of the unit. Out of eleven villages were selected seven villages from seven districts belonging to core regions of rice and four villages from four districts belonging to core regions of wheat. These rice and wheat villages are classified below.

6.2 DESIGN OF THE SAMPLE STUDY

The regional framework presented in Chapter 4, provided the basis for sample survey. From each village, eighteen households were interviewed, six each from large, medium and small holding sizes respectively. In all 198 landholdings from among 11 villages were selected for generating the primary data. A detailed questionnaire was filled for each household. The households within each village were selected on the basis of

TABLE 6.1 **Classification of Rice and Wheat Villages**

S.No.	Region	Crop	Village	Block	District	State
1	2	3	4	5	6	7
1.	West Coast	Rice	Uppoor	Udipi	South Kanara	Karnataka
2.	West Coast	Rice	Kannambra	Alatur	Palghat	Kerala
3.	East Coast	Rice	Gotala	Gopal-puram	West Godavari	Andhra Pradesh
4.	East Coast	Rice	Kaduvan Gudy	Nanilam	Tajnore	Tamil Nadu
5.	East Coast	Rice	Tanchanna	Biridi	Cuttack	Orissa
6.	North-Eastern Peninsular Plateau	Rice	Hanhat	Kuru	Ranchi	Bihar
7.	Ganges Delta	Rice	Ban Bahadurpur	Diamond Harbour	Diamond Harbour	West Bengal
8.	U.P. (Western Plain)	Wheat	Tarauli Samoli	Chauva Han	Mathura	Uttar Pradesh
9.	Malwa Plateau	Wheat	Rajowa	Sagar	Sagar	Madhya Pradesh
10.	Central Irrigated Area of Haryana	Wheat	Bandrana	Pundri	Central Karnal	Haryana
11.	Punjab Plains	Wheat	Latala	Pakhobar	Ludhiana	Punjab

purposive sampling. Households were classified as large, medium and small on the basis of gross income (income from agriculture plus non-agricultural income). Gross Income was considered a better standard than size of holding because in an inter-regional analysis of responses of crops to new technology, it will be arbitrary to equate land holding of a particular size in one region with that in another region. For example, *Latala* in Ludhiana district and *Rajowa* in Sagar district of Madhya Pradesh would really offer a contrast.

6.3 NATURE OF DATA GENERATED

On the basis of the information collected in the questionnaire data pertaining to twenty four variables were generated for the reference year 1975-76 for each village. These variables relate to the following :

TABLE 6.2 **List of Variables**[2]

1. Size of Holding.
2. Gross Cropped Area.
3. Net Area Sown.
4. Gross Cropped Area Irrigated.
5. Gross Cropped Area under High Yielding Variety.
6. Total Area under Rice and Wheat.
7. Total Area vnder High Yielding Variety of Rice and Wheat.
8. Total Area Irrigated of Rice.
9. Yield Per Hectare (in kg) of Irrigated Rice and Wheat.
10. Yield Per Hectare (in kg) of Unirrigated Rice and Wheat.
11. Yield Per Hectare (in rupees) of Irrigated Rice & Wheat.
12. Yield Per Hectare (in rupees) of Unirrigated Rice and Wheat.
13. Total Cost of Inputs (Including Land Revenue (in rupees) Labour and Miscellaneous Cultivation Expenses.
14. Total Cost of Inputs only (Seeds, Manure, all crops) and Inorganic Manures like Fertilizers.
15. Cost of Inputs used in Rice and Wheat.
16. Cost of Fertilizer Inputs in Rice and Wheat.
17. Total Agricultural Income.
18. Total Agricultural and Non-agricultural Income.
19. Net Agricultural Income.
20. Value of Rice and Wheat Output.
21. Totale Marketed Value (all crops).
22. Marketed Value of Rice and Wheat.
23. Value of Farm and Irrigated Equipment.
24. Value of Livestock.

6.4 ANALYSIS OF PRIMARY DATA

An attempt is made here to draw generalisations based on the observed pattern to see whether there is any nearness in the association between different landholdings in terms of certain chosen variables. For example, would it be true to say that the large holdings in one region are really similar to medium or small holdings in another region, in their adoption of modern agricultural practices. Large holdings in some rice regions may be associated with the small and medium holdings in a progrxssive wheat region in their responses to factors that influence crop productivity. Therefore, apart from bringing out inter-regional variations in rice and wheat responses to modern-methods in cultivation, the analysis that follows also cuts through vertically between different sizes of landholdings to bring out inter-landholding comparisons within and between regions.

Analysis of primary data has been attempted at three stages. Firstly, the raw data have been converted to percentages to facilitate the analysis of inter-regional and intra-regional holdings (Table 6 3). Secondly, pattern of association among different variables has been analysed with the help of correlation matrix first, for all samples of rice and wheat regions taken together and then, for large, medium and small categories of holdings. Thirdly, these variables were then sorted out and grouped into factors on the basis of factor loadings. This formed the basis for grouping 198 landholdings into six clusters. Similarities and dissimilarities among landholdings within and between rice/wheat regions served to identify the factors associated with regional variations in the responses of rice and wheat to new technology which is the central theme of this investigation.

6.5 INTRA-REGIONAL VARIATIONS

Tables 6.4, 6.5, 6.6 and 6.7 bring out the variations among rice regions and Tables 6.8, 6.9 and 6.10 give the same information for wheat regions. The analysis is presented first for rice regions and then for wheat regions.

TABLE 6.3 Transformed Variables

The variables listed in Table 6.2 were transformed into as given below :

1. Crop intensity (percentage of gross cropped area to net area sown).
2. Gross area irrigated as a percentage of gross cropped area.
3. Percentage of gross area under High Yielding Variety to gross cropped area.
4. Area under rice and wheat as a percentage of gross cropped area.
5. Rice and wheat irrigated area as a percentage of total area under rice/wheat.
6. Area under High Yielding Variety of rice/wheat to area under rice/wheat.
7. Percentage of total agricultural costs (excluding operational costs such as labour, miscellaneous expenses, and land revenue including irrigation cess) to total agricultural operational costs.
8. Cost of rice and wheat inputs as a percentage of total operational costs.
9. Cost of fertilizers used in rice and wheat as a percentage of total rice and wheat inputs costs.
10. Value of total rice/wheat output as a percentage of value of total agricultural output produced.
11. Value of output of rice/wheat marketed as a percentage of value of total output marketed.
12. Value of total output of rice and wheat marketed as a percentage of value of total rice and wheat output produced.
13. Income from agriculture as a percentage of gross income from all sources including non-agricultural income.
14. Gross agricultural income per hectare of gross cropped area.
15. Total agricultural operational costs per hectare of gross cropped area.
16. Total input costs per hectare of gross cropped area.
17. Rice and wheat input costs per hectare of total area under rice and wheat.
18. Value of fertilizers used in rice and wheat per hectare of total area under rice and wheat.

Rice Regions

Table 6.4 : This brings out some salient features of crop pattern and productivity in the West Coast region. In the villages belonging to West Coast region, the average size of large holdings is 3.90 hectares, of medium sized holdings is 1.82 hectares and that of small holdings is 1.05 hectares. In this region area under rice as a percentage of area under all crops

TABLE 6.4 **Regional Percentage of Different Transformed Variables by Size of Landholdings (Aggregates of Different Landholdings within the *West Cost Region*)**

Variables	Large (3.90 Hectares)	Medium (1.82 Hectares)	Small (1.05 Hectares)
1.	146.00	185.00	185.00
2.	95.30	95.50	93.80
3.	54.00	48.00	27.00
4.	91.00	92.15	95.20
5.	100.00	100.00	100.00
6.	43.00	52.00	28.40
7.	62.80	61.00	83.70
8.	53.90	55.40	74.60
9.	29.50	19.20	28.60
10.	73.20	84.08	88.39
11.	70.20	79.70	80.80
12.	82.10	70.70	60.16
13.	86.90	69.00	85.90
14.	4648.90	4641.78	4437.43
15.	1689.00	1887.00	1341.53
16.	1060.03	1157.00	1123.00
17.	1034.00	1136.76	1050.26
18.	305.00	218.90	301.54

Note : The list of variables appears before the tables in Table 6.3 of this chapter.

is 91.90 per cent for large holdings, 92.15 per cent for medium and 95.20 per cent for small holdings of land. This clearly brings out the fact that there are some minor variations in small and medium sized holdings as compared to the large holdings. However rice dominated the cropped area. Crop intensity is 146 per cent in the large holdings, 185 per cent in the case of medium and small holdings. Intensity is quite high in all the three categories of holdings though quite interestingly the survey reveals the fact that intensity of cultivation is more in small sized holdings. This is where the new farm strategy has to make a breakthrough keeping in view the dominance of small holdings in the region. Cultivation is restricted to mostly rice in all the three seasons. This is due to the fact that within the region almost entire cropped area under rice is rainfed during Kharif season and irrigated by tanks and wells during rabi season and inter-season variety sown between Kharif and Rabi.

In the adoption of high yielding varieties the region does not seem to have made much progress in contrast to the major wheat producing regions where apart from wheat, rice is essentially grown as a commercial crop by full use of modern technology. In 1975-76 only 43 per cent of total rice area was under high yielding variety in the case of large holdings while it was higher (52 per cent of total area under rice) in medium holdings. For small holdings, it was only 28 per cent. The average annual returns from agriculture for the three categories of holdings were Rs. 4649.00 per hectare of gross cropped area for large holdings, Rs. 4642.00 for medium holdings and Rs. 4437.00 for small holdings. There is not much variation in the total returns from agriculture even though the small landholders have not taken up to the use of high yielding variety. In fact in some areas, the traditional variety using greenleaf and cowdung as manure seemed to compare well with the returns by the use of modern inputs under the existing level of agricultural development in the region. Compared to this, the average operational costs of agriculture per hectare of gross cropped area is higher for medium holdings (Rs. 1887.00), as compared to large (Rs. 1689.00). For small holdings it is Rs. 1341.00. Operational costs of agriculture for small holdings are found to be lower compared to the returns as between large versus medium and small holdings who pay more attention to the land. Out of the input cost, fertiliser input amounts for 29.50 per cent in the case of large, 19.20 per cent in medium and 28.60 per cent in case of small holdings.

It will be noticed, that in this region, so far as the adoption of high yielding variety is concerned, in terms of area of rice under high yielding variety the small holdings have a lower percentage. But their input expenses on rice cultivation as a percentage of total operational costs is the highest. Rice input costs per hectare are also not very much lower than for the high and medium categories. This does point to the fact that the cost of rice cultivation is rather high for the small holdings of this region. In fact a balance has to be struck between traditional variety and its returns by intensive cultivation vis-a-vis the use of HYV by the application of inorganic fertilisers. The value of percentage of marketed surplus to total output of rice is of the order of 82 per cent for large, 71 per cent for medium and 60 per cent for small holdings. The percentage is rather high. This could be due to the lower level of consumption of the farmers in general in the region and so a low percentage is retained for self consumption. The tendency is to sell away a large part of the produce soon after harvest and then to buy the requirements as and when needed.

West Coast Region

Table 6.5 : This brings out the relative position of large, medium and small holdings in the East Coast region comprising villages of Gotala (West Godavari), *Kaduvangudv* (Tanjore) and Tanchanna (Cuttack). The average size of holding is 8.56 hectares for large, 1.88 hectares for medium and 1.04 hectares for small holdings. The cropping pattern in this region shows a high percentage of gross cropped area under rice. It is 76 per cent for the large holdings and 74 per cent for small holdings. In the case of medium sized holdings only 26 per cent of total area is under rice. But the intensity of cropping for medium holdings is 209 per cent. This shows that the cropping pattern in this size of land holding is highly diversified.

In general intensity of cropping in the region is high for all the holding sizes. The percentages of rice irrigated area to area under rice is 80.80 per cent for large, 76.68 per cent for medium and 48.22 per cent for small holdings. This is in contrast to West Coast region where holdings are generally small in size and intensively cultivated.

TABLE 6.5 **Regional Percentage by Size of Landholdings of Different Variables (Aggregates of Different Landholdings within the *East Cost Region*)**

Variables	Large (8.56 Hectares)	Medium (1.88 Hectares)	Small (1.04 Hectares)
1.	161.00	209.00	182.00
2.	80.00	68.60	37.60
3.	46.90	61.20	27.50
4.	76.10	26.00	74.20
5.	80.80	76.68	48.22
6.	61.49	71.90	37.05
7.	38.88	58.17	46.98
8.	35.95	49.98	46.11
9.	49.39	51.80	50.80
10.	74.40	93.30	94.80
11.	58.49	96.70	91.46
12.	79.00	65.90	45.80
13.	97.20	93.10	80.40
14.	4057.00	3521.00	2016.00
15.	1400.00	1331.00	862.00
16.	575.00	681.00	405.00
17.	698.00	781.00	535.00
18.	347.00	405.00	272.00

So far as the area under high yielding variety is concerned in the case of large holdings, 61.49 per cent of total area is under rice. In the medium and small holdings it is 71.90 per cent and 37.05 per cent respectively. The medium holdings have the highest cropping intensity and highest area under high yielding variety as compared to the other two categories. This is corroborated by the fact that the percentage of operational costs of agriculture and the percentage of rice input costs to total operational costs is also the highest for medium landholdings. However, the cost of inputs as a percentage of total operational costs, and the cost of fertilizer input as a percentage of input costs on rice cultivation are almost equally high for the large as well as small holdings. The gross agricultural income per hectare of gross cropped area is the highest for large holdings followed by medium holdings while it is relatively very low for small holdings. The total operational costs of cultivation per hectare of gross cropped area are Rs. 1480.00 for large and Rs. 1331.00 for medium and Rs. 862.00 for small holdings. But the fertilizers input costs per hectare of rice area and total input costs per hectare of rice area are almost the same for the three holding sizes. This goes to show that even in the small holdings cost of rice cultivation is rather high and the returns relatively lower. The marketed surplus is the highest in the large holdings followed by medium and small holdings. For example, the percentage value of rice output sold to total output of rice is 79 for large, 66 for medium and 46 for small holdings. This partly brings out the difference in commercialisation of rice cultivation which is restricted more to large holdings. In some cases it is also possibly due to cash payment towards labour and other operational expenses which compel the farmers to sell large part of the produce.

North Eastern Peninsular Plateau Region

Table 6.6 : This brings out the salient features of cropping pattern and crop productivity between the holdings of different sizes in the north eastern peninsular plateau, from which village (Hanhat-Chota Nagpur area) in Ranchi district was selected. The average size of holdings in the large holding category is 6.53 hectares, 3.26 hectares in the medium category and 1.59 hectares in the small holdings. The intensity of cropping is 149 per cent in the case of large holdings, 142 per cent in case of medium and 134 per cent in case of small holdings. The area under rice as a percentage of gross cropped area is 43.30 per cent for large holdings, 55.20 per cent for medium and 70.27 per cent for small holdings. This observation explains two inherent characteristics associated with crop rotation and influence of size of landholding. Firstly large holdings make use of available irrigation for rice-wheat rotation and hence the percentage of cropped area under rice decreases. Secondly

TABLE 6.6 **Regional Percentage of Different Variables by Size of Landholdings (Aggregates of Different Landholdings within the *North Eastern Peninsular Plateau Region*)**

Variables	Large (6.53 Hectares)	Medium (3.26 Hectares)	Small (1.59 Hectares)
1.	149.00	142.00	134.00
2.	57.00	56.70	61.20
3.	23.40	22.66	25.60
4.	43.30	55.20	70.27
5.	47.60	63.20	50.00
6.	—	—	—
7.	84.30	90.69	90.69
8.	14.85	33.98	28.00
9.	30.00	49.40	24.08
10.	23.19	52.90	59.20
11.	32.70	37.15	40.25
12.	45.50	40.92	22.86
13.	87.80	59.67	72.05
14.	2898.00	1669.00	2339.00
15.	1902.00	1229.00	1115.00
16.	1605 00	1115.00	846.00
17.	651.00	756.00	458.00
18.	191.00	374.00	110.00

small holdings are characterised by rice-pulses rotation, the latter crop requiring less irrigation than required for wheat. Area devoted to high yielding variety of rice is very little or nil. This is to be expected because in the kharif season, rice cultivation depends on the timely arrival of rains. Therefore, there is lot of risk involved in the cultivation of high yielding varieties in the absence of assured and sufficient water supply. There is also the problem of diffusion of innovations which have reached Western U.P. from Punjab-Haryana area while Bihar is situated in the midst of eastern U.P. and West Bengal where the use of H.Y.V. of rice is still in its initial stage of adoption. Socio-economic and institutional factors are in impediment for modernisation of agriculture in this area. Irrigated area under rice as a percentage of area under rice is 43.30 per cent for large holdings, 63.20 per cent for medium and 50.00 per cent for small holdings respecitvely.

Ganges Delta Region

Table 6.7 : This relates to the Ganges Delta region. Village *Banbahadurpur* of 24 parganas district was selected. The average size of landholding in the region is 3.84 hectares in the case of large, 1.86 hectares in the case of medium and 0.84 hectare in the case of small holdings. Due to high rainfall, the cultivated area is mostly rainfed in the kharif season. Intensity of cropping in the case of large holdings reported to be nil and it is very low for medium (2%) and small holdings (1%). In the study area there is only one crop of rice during the whole year. Rice accounts for more than 97 per cent of total crop area in all the three categories of holding sizes. The input costs for rice as a percentage of total operational costs is 39.82 per cent for large holdings, 53.89 per cent for medium and 51.00 per cent for small holdings. The small holdings have highest total operational costs per hectare. Their contribution to marketed output of rice is nil. The yields are so low that the production is hardly sufficient for self consumption. In fact the farmers purchase additional rice from market, in order to meet the consumption requirements. The small holdings do not use fertilisers at all, and the large and medium holdings provide to the extent of 32.05 per cent and 12.15 per cent of the total input cost to the fertilizers. Also in this village, the small holdings derive only 40.34 per cent of income from agriculture. The rest of it is accounted for, by non-agricultural sources of income, mainly sale of fish.

TABLE 6.7 **Regional Percentage of Different Variables by Size of Landholdings (Aggregates of Different Landholdings within the *Ganges Delta Region*)**

Variables	Large (3.84 Hectares)	Medium (1.86 Hectares)	Small (0.84 Hectare)
1.	100.00	102.00	101.00
2.	100.00	100.00	100.00
3.	—	—	—
4.	100.00	97.38	97.70
5.	100.00	100.00	100.00
6.	—	—	—
7.	39.82	53.89	51.00
8.	39.82	53.60	50.90
9.	32.05	12.15	—
10.	100.00	98.19	82.22
11.	100.00	86.70	—
12.	40.66	15.25	—
13.	69.82	69.26	40.34
14.	2497.00	2255.00	3240.00
15.	917.00	1043.00	516.00
16.	365.00	624.00	263.00
17.	365.00	568.00	263.00
18.	117.00	71.00	—

Wheat Region

Western U.P. Plains

Table 6.8 : The average size of holding in the region is7.89 hectares for large holdings, 3.37 hectares for medium and 1.21 for small holdings. Intensity of cropping is higher in the case of medium holdings. The percentage of gross area irrigated to gross cropped area is 75.11 per cent for large, 68.86 per cent for medium and 84.20 per cent for small holdings. In fact much of irrigation water is used for sugarcane cultivation for

TABLE 6.8 **Regional Percentage of Different Variables by Size of Landholdings (Aggregates of Different Landholdings within the *U.P. Western Plains Region*)**

Variables	Large (7.89 Hectares)	Medium (3.37 Hectares)	Small (1.21 Hectares)
1.	151.00	180.00	142.00
2.	75.11	68.86	84.20
3.	33.50	33.33	36.84
4.	34.51	34.40	36.80
5.	100.00	100.00	100.00
6.	88.26	90.31	100.00
7.	63.98	74.47	54.00
8.	24.97	33.80	26.20
9.	50.90	28.50	61.60
10.	39.50	36.90	38.00
11.	26.88	33.41	27.63
12.	43.10	54.00	57.60
13.	100.00	87.50	81.70
14.	2394.00	2767.00	4321.00
15.	1047.70	1467.70	995.80
16.	1005.00	538.60	237.00
17.	1442.00	708.80	338.00
18.	386.00	411.66	437.36

which Western U.P. is well-known. The area under high yielding varieties of wheat as percentage of total cropped area is 88.26 per cent in the case of large holdings, 90.31 per cent in the case of medium and 100 per cent in the case of small holdings. This is due to the fact that some of the farmers in the large and medium category of holdings cultivate rice or sugarcane on a small area and therefore, the area under high yielding variety of wheat is lower in the case of large and medium holdings. The cost of inputs for wheat as a percentage of total operational costs is 24.97 per cent for large, 33.80 per cent for medium and 26.20 per cent for small holdings. This is very low considering the fact that a high percentage of wheat area is under high yielding variety. Operational costs are obviously high involving the use of modern agricultural implements and hired labour. However, the value of fertilizers used as a percentage of wheat input costs is high. It is 50.90 per cent for large, 28.50 per cent for medium and 61.60 per cent for small holdings. For the medium holdings considering the fact that a very high percentage of area is under wheat the proportion of fertilizer cost to total input cost on wheat area is very low. The marketed surplus of wheat is 43.10 per cent of total wheat output for large holdings, 54 per cent for medium holdings and 57.60 per cent for small holdings. In all the three categories almost half of the output is retained for self consumption.

Malwa Plateau Region

Table 6.9 : The average size of holding in the study area is 28.12 hectares for large, 7.51 hectares for medium and 1.97 for small holdings. The acreages under all sizes of holdings is more in comparison with those in Western U.P. Much of the land belongs to dry undulating topography developed over the basaltic rocks of this part of the peninsular plateau. This is corroborated by the findings relating to crop intensity and other characteristics. The intensity of cropping is negligible in the case of holdings of all sizes. This is evident from the fact that much of the land is unirrigated. The area under high yielding variety of wheat is 51.72 per cent of the total area under wheat for large holdings and nil for the other two categories. Fertilizer costs per hectare of wheat area is extremely low for medium and small categories. The marketed surplus of wheat is 56.20 per cent of total wheat output in the high holding category and negligible for medium and small holdings. This is due to the low yields, especially among the medium and small holdings. In fact in the small holdings most of the farmers depend for their living on income from agricultural sources. The most important source of extra income is Beedi making. They are generally paid in kind for this work.

Punjab Plains

Table 6.10 : This brings out the salient features of cropping

TABLE 6.9 **Regional Percentage of Different Variables by Size of Landholdings (Aggregates of Different Landholdings within the *Malwa Plateau Region*)**

Variables	Large (28.12 Hectares)	Medium (7.5 Hectares)	Small (1.97 Hectares)
1.	113.00	109.00	100.00
2.	40.20	28.68	0.09
3.	26.97	—	—
4.	52.15	59.60	83.70
5.	51.72	27.00	0.06
6.	51.72	—	—
7.	51.27	76.23	75.76
8.	36.56	39.33	65.17
9.	19.67	8.63	2.10
10.	75.70	43.90	74.28
11.	78.60	1.85	—
12.	56.20	1.47	—
13.	95.31	100.00	59.50
14.	1591.00	1287.00	551.00
15.	462.07	859.90	398.99
16.	237.00	654.00	327.00
17.	338.00	338.70	0.59
18.	210.00	0.19	0.05

pattern and crop productivity relationship for this region. The average size of holding in the region is 9.9 hectares in the case of large holdings, 3.15 hectares in the case of medium and 1.24 hectares in case of small holdings. The intensity of cropping is high in the region as there is enough water from irrigation to grow crops in more than one season. The irrigated area of wheat as a percentage of total area under wheat is very high. Average area under high yielding variety is 88.61 per cent for large and 56.20 per cent for medium and small holdings. In fact the use of HYV was restricted to large holdings initially and is now extended to other categories as well. The cost of inputs for wheat cultivation as a percentage of total operational costs is almost the same for all the three categories. In fact input cost of fertilizer accounts for more than 60 per cent for all the

TABLE 6.10 **Regional Percentage of Different Variables by Size of Landholdings (Aggregates of Different Landholdings within the *Punjab Plains Region*)**

Variables	Large (9.9 Hectares)	Medium (3.15 Hectares)	Small (1.24 Hectares)
1.	159.00	148.00	144.00
2.	94.80	93.04	93.33
3.	69.33	72.94	84.00
4.	57.42	56.20	56.20
5.	100.00	100.00	100.00
6.	88.61	56.20	56.20
7.	76.17	76.17	72.16
8.	46.06	57.02	52.28
9.	61.95	60.20	68.14
10.	70.07	76.41	80.12
11.	68.02	47.15	80.63
12.	82.70	73.70	82 04
13.	84.56	83.68	89.18
14.	2444.00	2300.00	2516.00
15.	1473.00	1286.00	1391.00
16.	1122.00	986.00	1004.00
17.	1182.00	1302.00	1097.00
18.	732.00	784.00	747.00

three holding sizes. A striking feature here is that the variations in wheat cultivation in respect of modernisation are less marked among holdings of various sizes. This is due to the spread of modern methods of cultivation over most parts of the region. Green revolution was started in Punjab as early as in 1966 with relatively high level of agricultural development even under traditional methods of farming. Thus, most of the farmers have adopted the modern agricultural practices. The marketed samples of wheat as a percentage of total wheat output is as high as 82.70 per cent for large holdings, 73.70 per cent for medium holdings and 82.04 per cent for small holdings. Regional specialisation of agriculture finds its expression at various levels starting from the small holdings unlike in other regions where the impact of Green Revolution has yet to percolate down to the small farmers.

6.6 INTER REGIONAL ANALYSIS

When we compare the different sizes of landholdings and the results obtained for them between regions, the following features emerge from the above analysis.

Firstly, the average size of holdings in the rice region is almost the same specially for the medium and small categories. However the average size of holding in the large category is much higher than for example, in the West Coast Region. Cropping intensity is high in all the rice regions.

When all the rice regions are considered, the process of adoption of high yielding variety seems to be still slow even as late as in 1975-76. However in wheat regions by contrast, a very high percentage of total area of wheat is under H.Y.V. The total operational costs per hectare of total cropped area is higher in the rice regions than in the wheat regions. This is due to the fact that in the West Coast region and some parts of East Coast region, rice is cultivated during all the three seasons and therefore per hectare costs may appear higher, but between wheat regions and rice regions they are not very different. The only exception to this is the Malwa Plateau of Madhya Pradesh where the cost of wheat cultivation is rather low due to traditional methods of cultivation and cultivation of traditional variety of wheat especially by medium and small farmers and poor quality of the land. Secondly, marketed surplus of rice is higher for medium and small categories of holding in the rice regions than in the wheat regions. This could be due to the lower level of consumption of farmers in rice regions and also due to the fact that in these regions joint family system is not prevalent on a large scale whereas in the wheat villages many households do have large numbers of dependents to support as shown in the sample surveys. Hence, a large amount of output is retained for self consumption. The only exception to this is Punjab-Haryana region where marketed surplus of wheat is rather high. Finally, input costs for rice and wheat as a percentage of total operational costs are almost the same between rice regions and wheat regions. Exception to this is the Malwa Plateau region where the input costs are very low due to traditional methods of cultivation. This is directly connected with lack of irrigation facilities in the region.

6.7 RESULTS OF MULTIPLE CORRELATION

In order to analyse the relationship between the variables multiple correlation was attempted. The co-efficient of correlation is given for all the twenty four variables considered in this analysis in Tables 6.11, 6.12, 6.13 and 6.14. First a correlation matrix has been presented for all samples of large, medium and small holdings together. This is then followed by a separate correlation matrix for the three categories of holdings.

All Samples

First, we will take up the analysis of the correlation between area under rice and wheat and some other variables connected with crop acreage. These are (i) gross area irrigated, (ii) area irrigated of the crop, (iii) yield in rupees per hectare and gross cropped area. The co-efficient of correlation between area under rice and wheat and gross area irrigated is very high (0.92). This is of course a well-known fact. There is a very low correlation between area under the crop and yield (in rupees per hectare). Finally, between gross cropped area and area under the crop, there is a high correlation (0.92). Irrigated area of rice/wheat shows positive correlation with input costs (0.44), total agricultural income (0.41), net agricultural income (0.49), total income from rice output (0.53) and gross income from all sources (0.48). The association between area under high yielding variety of rice/wheat is highly positive with the following variables :

i. Cost of inputs for the crop.
ii. Value of fertilizers used in rice/wheat (0.73).
iii. Total agricultural income (0.67).

iv. Net agricultural income (0.71).
v. Total value of rice/wheat output (0.84).

This shows that adoption of high yielding variety is closely associated with the input costs and income derived from agriculture as a whole. To somewhat lesser extent, the area under high yielding variety of the crop is also correlated with the value of farm and irrigation equipment (0.42) and value of livestock (0.41). In other words, there is a high correlation between size of holdings and cost of inputs for rice/wheat (0.89).

There is a high correlation between size of holdings and value of farm and irrigation equipment (0.95) and value of livestock (0.87). It can be concluded from this that size of holdings and value of farm and irrigation equipment are interrelated.

Large Samples

The co-efficient of correlation between area under rice/wheat and value of gross area irrigated is 0.35. But there is high correlation between area under the crop and area irrigated of the crop (0.81). Between unirrigated rice and wheat yields and area under them the correlation is negative. There is a high correlation between area under the crop and gross cropped area (0.82). Taking yield per hectare (in kilograms) of irrigated rice and wheat it is found that the correlation values are low. However, there is positive correlation between yield and net agricultural income (0.37), total income from rice and wheat output (0.45). This is in contrast to results obtained for all samples taken together. Between unirrigated rice and wheat yield per hectare in kilograms and gross income from agricultural sources, the value of correlation is 0.33. This only goes to prove that farmers with irrigated area under rice and wheat even in large holding category have a high percentage of income from non-agricultural sources. There are large holdings of large size but they are not productive in some areas e.g. in parts of Malwa Plateau for wheat.

It will be noticed that there is a negative correlation between size of holding and yield per hectare. There is fairly high correlation between size of holding and total agricultural

TABLE 6.11 **Correlation Matrix—All Samples of Holdings**[2]

Variables	1	2	3	4	5	6	7	8	9	10	11	12
1.	1.00											
2.	0.86	1.00										
3.	0.90	0.95	1.00									
4.	0.66	0.87	0.75	1.00								
5.	0.48	0.58	0.48	0.66	1.00							
6.	—0.01	0.92	0.88	0.84	0.63	1.00						
7.	—0.04	0.56	0.47	0.62	0.95	0.65	1.00					
8.	0.03	0.01	0.01	0.02	0.02	0.01	0.02	1.00				
9.	0.03	0.14	0.04	0.28	0.36	0.25	0.40	0.04	1.00			
10.	0.03	0.04	0.03	0.05	0.13	0.10	0.10	—0.05	0.09	1.00		
11.	0.59	0.80	0.05	0.27	0.37	0.24	0.41	0.06	0.99	0.06	1.00	
12.	0.42	0.62	0.03	0.04	0.13	0.10	0.11	0.05	0.08	0.99	0.06	1.00
13.	0.33	0.54	0.08	0.85	0.57	0.78	0.53	0.02	0.24	0.88	0.25	0.09
14.	0.20	0.45	0.49	0.73	0.67	0.56	0.22	0.00	0.22	0.00	0.24	0.00
15.	0.89	0.83	0.40	0.68	0.79	0.64	0.77	0.01	0.44	—0.01	0.45	—0.01
16.	0.48	0.60	0.30	0.63	0.77	0.51	0.73	0.01	0.32	—0.12	0.33	—0.13

17.	0.53	0.60	0.73	0.85	0.68	0.86	0.67	0.09	0.41	0.13	0.41	0.13
18.	0.58	0.63	0.53	0.58	0.69	0.66	0.71	0.00	0.49	0.11	0.48	0.10
19.	0.46	0.58	0.49	0.64	0.79	0.73	0.84	0.01	0.53	0.06	0.52	0.54
20.	0.57	0.69	0.52	0.67	0.70	0.67	0.67	0.02	0.48	0.14	0.48	0.13
21.	0.38	0.44	0.55	0.57	0.13	0.60	0.13	0.02	0.11	0.08	0.11	0.09
22.	0.40	0.47	0.63	0.69	0.35	0.76	0.37	0.02	0.24	0.09	0.24	0.09
23.	0.95	0.44	0.42	0.44	0.48	0.41	0.42	0.01	0.12	0.12	0.13	0.13
24.	0.87	0.47	3.43	0.50	0.43	0.38	0.41	0.10	0.09	0.10	0.12	0.10

TABLE 6.11 (*Contd.*)

Variables	13	14	15	16	17	18	19	20	21	22	23	24
1.												
2.												
3.												
4.												
5.												
6.												
7.												
8.												
9.												
10.												
11.												
12.												
13.	1.00											
14.	0.81	1.00										
15.	0.66	0.78	1.00									

16.	0.57	0.69	0.88	1.00								
17.	0.88	0.72	0.69	0.53	1.00							
18.	0.43	0.48	0.62	0.43	0.79	1.00						
19.	0.52	0.53	0.80	0.64	0.77	0.88	1.00					
20.	0.59	0.66	0.73	0.57	0.79	0.86	0.85	1.00				
21.	0.71	0.28	0.17	0.10	0.62	0.14	0.14	0.12	1.00			
22.	0.76	0.36	0.37	0.26	0.75	0.36	0.42	0.35	0.92	1.00		
23.	0.40	0.50	0.42	0.35	0.47	0.40	0.37	0.50	0.07	0.19	1.00	
24.	0.40	0.42	0.35	0.36	0.41	0.31	0.29	0.34	0.15	0.21	0.29	1.00

1. See Section 6-3 of this chapter for the List of Variables.
2. Number of Samples 198.

TABLE 6.12 **Correlation Matrix—Large Holdings**

Variables	1	2	3	4	5	6	7	8	9	10	11	12
1.	1.00											
2.	0.80	1.00										
3.	0.86	0.93	1.00									
4.	0.53	0.82	0.65	1.00								
5.	0.30	0.32	0.23	0.38	1.00							
6.	0.70	0.88	0.83	0.77	0.40	1.00						
7.	0.35	0.34	0.25	0.35	0.92	0.46	1.00					
8.	0.50	0.66	0.50	0.82	0.60	0.81	0.63	1.00				
9.	—0.17	—0.14	—0.20	—0.07	0.15	0.07	0.25	0.20	1.00			
10.	—0.10	—0.16	—0.11	—0.24	—0.33	—0.05	—0.25	—0.19	0.35	1.00		
11.	—0.18	—0.12	—0.18	—0.07	0.17	0.07	0.27	0.20	0.98	0.29	1.00	
12.	0.12	—0.16	—0.11	—0.27	—0.34	—0.05	—0.27	—0.19	0.35	0.99	0.30	1.00
13.	0.37	0.66	0.51	0.74	0.25	0.63	0.22	0.53	—0.06	—0.03	—0.04	—0.02
14.	0.06	0.34	0.18	0.49	0.45	0.19	0.30	0.29	—0.13	—0.00	—0.08	—0.19

15.	0.03	0.19	0.01	0.40	0.68	0.33	0.66	0.58	0.31	—0.19	0.35	—0.18
16.	0.10	0.16	0.03	0.40	0.63	0.21	0.58	0.46	0.14	—0.29	0.17	—0.29
17.	0.58	0.74	0.63	0.75	0.44	0.79	0.46	0.72	0.20	0.07	0.20	0.05
18.	0.43	0.37	0.33	0.31	0.56	0.45	0.60	0.54	0.37	0.02	0.36	0.01
19.	0.35	0.34	0.24	0.37	0.71	0.55	0.79	0.74	0.45	—0.06	0.43	—0.09
20.	0.35	0.31	0.19	0.35	0.48	0.36	0.48	0.50	0.38	0.83	0.37	0.06
21.	0.41	0.58	0.54	0.51	—0.05	0.65	—0.03	0.42	—0.00	0.07	—0.00	0.08
22.	0.50	0.65	0.59	0.64	0.12	0.77	0.17	0.59	0.09	0.06	0.08	0.06
23.	0.18	0.20	0.22	0.18	0.31	0.16	0.23	0.15	—0.01	0.10	0.01	0.11
24.	0.40	0.39	0.35	0.41	0.19	0.18	0.18	0.24	—0.28	0.16	—0.25	—0.16

TABLE 6.12 (*Contd.*)

Variables	13	14	15	16	17	18	19	20	21	22	23	24
1.												
2.												
3.												
4.												
5.												
6.												
7.												
8.												
9.												
10.												
11.												
12.												
13.	1.00											
14.	0.65	1.00										
15.	0.33	0.57	1.00									

16.	0.32	0.55	0.88	1.00								
17.	0.80	0.77	0.39	0.23	1.00							
18.	0.07	0.12	0.37	0.16	0.61	1.00						
19.	0.13	0.15	0.66	0.50	0.53	0.81	1.00					
20.	0.15	0.32	0.49	—0.32	0.51	—0.78	0.73	1.00				
21.	0.77	0.14	—0.02	—0.06	0.67	—0.05	0.05	0.14	1.00			
22.	0.75	0.14	0.12	0.02	0.74	0.12	0.19	0.14	0.93	1.00		
23.	0.13	0.30	0.17	0.16	0.25	0.30	0.14	0.31	—0.06	0.00	1.00	
24.	0.19	0.14	—0.01	0.02	0.23	0.16	0.06	0.08	0.08	0.08	0.07	1.00

1. Number of Samples 61.

TABLE 6.13 **Correlation Matrix—Medium Holdings**

Variables	1	2	3	4	5	6	7	8	9	10	11	12
1.	1.00											
2.	0.84	1.00										
3.	0.98	0.87	1.00									
4.	0.04	0.35	0.06	1.00								
5.	—0.07	0.23	—0.04	0.67	1.00							
6.	0.69	0.75	0.71	0.22	0.25	1.00						
7.	—0.11	0.24	0.06	0.65	0.95	0.32	1.00					
8.	—0.09	0.19	0.07	0.76	0.67	0.40	0.73	1.00				
9.	—0.47	—0.23	—0.45	0.24	0.45	—0.01	0.52	0.44	1.00			
10.	—0.05	0.00	—0.05	—0.09	—0.23	0.06	—0.20	—0.08	0.14	1.00		
11.	0.46	—0.21	—0.43	0.28	0.47	—0.07	0.54	—0.45	0.99	0.13	1.00	
12.	—0.04	0.01	—0.05	—0.09	—0.24	0.06	—0.21	—0.09	0.13	0.99	0.12	1.00
13.	0.24	0.45	0.21	0.69	0.48	0.30	0.47	0.53	0.19	0.08	0.22	0.09
14.	0.30	0.43	0.26	0.67	0.38	0.20	0.30	0.38	0.08	0.06	0.11	0.07
15.	0.12	0.32	0.12	0.70	0.60	0.42	0.60	0.73	0.43	0.43	0.03	0.44

16.	0.00	0.24	0.02	0.68	0.75	0.26	0.72	0.60	0.34	—0.19	0.36	—0.20
17.	0.06	0.36	0.07	0.51	0.57	0.38	0.63	0.50	0.61	0.07	0.62	0.07
18.	—0.05	0.20	—0.02	0.25	0.42	0.30	0.52	0.45	0.65	0.05	0.66	0.05
19.	—0.11	0.13	—0.09	0.37	0.57	0.44	0.67	0.65	0.73	0.00	0.73	0.00
20.	—0.02	0.21	—0.02	0.41	0.49	0.26	0.54	0.52	0.63	0.65	0.64	0.06
21.	—0.11	0.14	—0.10	0.58	0.53	0.25	0.56	0.60	0.67	0.16	0.68	0.16
22.	—0.21	0.40	—0.18	0.46	0.58	0.33	0.66	0.70	0.75	0.04	0.74	0.03
23.	0.05	0.24	0.09	0.39	0.23	0.14	0.24	0.27	0.04	0.08	—0.03	0.08
24.	0.25	0.40	0.30	0.41	0.33	0.11	0.30	0.13	0.08	0.16	—0.05	—0.05

TABLE 6.13 (*Contd.*)

Variables	13	14	15	16	17	18	19	20	21	22	23	24
1.												
2.												
3.												
4.												
5.												
6.												
7.												
8.												
9.												
10.												
11.												
12.												
13.	1.00											
14.	0.87	1.00										
15.	0.03	0.72	1.00									

16.	0.43	0.42	0.64	1.00								
17.	0.64	0.54	0.67	0.48	1.00							
18.	0.24	0.18	0.43	0.24	0.89	1.00						
19.	0.36	0.23	0.64	0.49	0.82	0.81	1.00					
20.	0.53	0.43	0.55	0.33	0.85	0.75	0.76	1.00				
21.	0.61	0.53	0.75	0.42	0.80	0.66	0.75	0.72	1.00			
22.	0.39	0.25	0.67	0.50	0.72	0.66	0.90	0.58	0.87	1 00		
23.	0.31	0.37	0.38	0.17	0.20	0.07	0.07	0.09	0.29	0.16	1.00	
24.	0.33	0.35	0.31	0.35	0.17	0.02	0.00	0.10	0.18	0.06	0.33	1.00

1. Size of Samples 62.

TABLE 6.14 **Correlation Matrix—Small Holdings**

Vari-ables	1	2	3	4	5	6	7	8	9	10	11	12
1.	1.00											
2.	0.33	1.00										
3.	0.46	0.92	1.00									
4.	0.01	0.80	0.74	1.00								
5.	0.00	0.78	0.74	0.91	1.00							
6.	0.23	0.91	0.86	0.77	0.77	1.00						
7.	—0.04	0.77	0.11	0.88	0.96	0.80	1.00					
8.	—0.06	—0.06	0.05	—0.02	—0.01	—0.08	0.00	1.00				
9.	—0.23	0.24	0.10	0.46	0.38	0.23	0.39	0.08	1.00			
10.	0.10	0.13	0.07	0.28	—0.19	0.10	—0.18	—0.07	—0.32	1.00		
11.	—0.22	0.25	0.11	0.47	0.40	0.22	0.41	0.10	0.99	0.34	1.00	
12.	0.10	0.13	0.06	0.29	0.19	0.86	—0.18	—0.07	—0.34	0.99	0.36	1.00
13.	0.05	0.88	0.79	0.91	0.89	0.86	0.90	0.00	0.33	—0.06	0.34	—0.06
14.	0.06	0.79	0.74	0.90	0.87	0.84	0.88	—0.02	0.43	—0.19	0.35	—0.19
15.	0.07	0.78	0.72	0.89	0.85	0.83	0.87	—0.02	0.38	—0.20	0.39	—0.21

16.	—0.03	0.79	0.74	0.91	0.90	0.79	0.89	0.00	0.31	—0.17	0.32	—0.17
17.	0.06	0.76	0.65	0.00	0.76	0.77	0.77	0.33	0.53	—0.07	0.55	—0.08
18.	0.09	0.59	0.47	0.89	0.55	0.63	0.57	—0.10	0.60	—0.03	0.60	—0.05
19.	0.06	0.79	0.67	0.81	0.77	0.83	0.79	—0.16	0.57	0.00	0.53	—0.01
20.	0.06	0.75	0.64	0.81	0.77	0.75	0.76	—0.08	0.49	—0.07	0.50	—0.08
21.	0.07	0.83	0.73	0.88	0.85	0.83	0.85	—0.04	0.47	—0.13	0.48	—0.14
22.	0.07	0.82	0.72	5.86	0.84	0.83	0.85	—0.05	0.47	—0.10	0.47	—0.11
23.	0.37	0.62	0.75	0.53	0.57	0.55	0.54	0.00	0.01	—0.16	0.03	—0.18
24.	0.02	0.53	0.51	0.58	0.64	0.53	0.64	0.19	0.18	—0.20	0.21	—0.20

TABLE 6.14 (*Contd.*)

Variables	13	14	15	16	17	18	19	20	21	22	23	24
1.												
2.												
3.												
4.												
5.												
6.												
7.												
8.												
9.												
10.												
11.												
12.												
13.	1.00											
14.	0.94	1.00										
15.	0.92	0.95	1.00									

16.	0.92	0.92	0.91	1.00								
17.	0.82	0.81	0.82	0.76	1.00							
18.	0.55	0.60	0.64	0.50	0.79	1.00						
19	0.82	0.80	0.87	0.77	0.89	0.87	1.00					
20.	0.80	0.80	0.80	0.76	0.83	0.81	0.89	1.00				
21.	0.89	0.89	0.88	0.83	0.88	0.79	0.91	0.88	1.00			
22.	0.86	0.84	0.88	0.82	0.86	0.79	0.94	0.87	0.97	1.00		
23.	0.54	0.55	0.52	0.58	0.38	0.19	0.36	0.37	0.50	0.47	1.00	
24.	0.59	0.64	0.50	0.63	0.54	0.29	0.39	0.44	0.54	0.47	0.56	1.00

1. Size of Samples 75.

income (0.58). This is in contrast to the pattern observed for all samples taken together where size of holding was highly correlated with a number of variables. On the whole for the large samples between area and factors affecting acreages, between yield and factors influencing yields the correlation values are not very high.

Medium Holdings

The area under rice and wheat have a low correlation between the factors affecting acreages. The correlation value is rather low, between area under the crop and area irrigated of the crop (0.40). The correlation between gross cropped area and area under the crop is 0.75. Between area and yield per hectare, there is a negative relationship. Certain interesting patterns emerge when we take up correlation values between yield per hectare of irrigated rice and wheat and other variable affecting yield. It will be noticed from correlation matrix 6.13, that there is a high correlation between yield per hectare and rice and wheat sale value (0.75), total sale value (0.67), total income from the crop (0.73), net agricultural income (0.61). It is this feature that was also reflected in the case of all samples. In this category of holding yields and returns from agriculture are highly correlated. There is however a negative correlation between size of holding and gross area under high yielding variety of rice and wheat and between area irrigated of rice and yield per hectare in the case of both irrigated and unirrigated rice and wheat. Further there is also a negative correlation between size of holding and net agricultural income, total crop income, gross income from agricultural and non-agricultural sources, total sale values of rice and wheat output. This clearly shows that yield levels are almost independent of the size of holding in the case of medium holdings. But yields and returns from agriculture are highly correlated. Area under high yielding variety is highly correlated with returns from agriculture.

Small Landholdings

For these samples, there is a negative correlation between area under rice and wheat and irrigated area of the crop. There is also a very low correlation between gross cropped area and area under the crop. This is due to the fact that intensity is very low for most of the small holdings. In fact, this is the reason why net area sown is highly correlated with a number of varibla-es in this category. This is evident from column 3 of Table 6.14. It is highly correlated with all the variables from (13-24) with some exceptions. The area under rice and wheat is highly corre-lated with total operational costs (0.86), cost of crop inputs (0.83), value of fertilizers used in the cultivation of the crop (0.79) and all other variables as is evident from column 6 of Table 6.14 (variables 17-24).

As in medium sized holdings for small holdings, also, size of holding is not highly correlated with most variables. Area under the crop however is highly correlated with returns from agriculture. There is also a high correlation between area under high yielding variety and returns from agriculture.

From the above analysis, it can be said that in the case of small holdings of land, returns from agricultural production are highly correlated with area under the crop—a sign of extensive as opposed to intensive cultivation. In the case of medium cate-gory holdings, the returns are highly correlated with yields. For large holdings size of holding and yield are highly correlated.

Another interesting result is that in the case of large holdings, the extent of correlation between area under high yielding variety costs and returns from agriculture is not that high as for medium and small holdings. It is the highest for small holdings. This proves that for medium and small farmers, adoption of high yielding variety will take into account factors affecting costs and returns from agriculture. In fact this is corroborated by our earlier analysis of regional percentages, where in some regions, the intensity of cropping is higher for medium landholdings and in few cases for small holdings. Therefore, a conclusion that can be drawn from such an analysis is that intensity in cultivation is highly correlated with total costs of and returns from cultivation.

6.9 CLUSTERS OF LANDHOLDINGS AND THE EMERGING PATTERNS OF ASSOCIATION

Factor analysis was used as the basis to separate out the variables with highest factor loadings on factors. This was worked out for all the 24 variables. This is given below.

TABLE 6. 15 **Highest Factor Loadings on Factors**

Variables	All Samples	Large	Medium	Small
1.	1	1	2	2
2.	1	1	2	1
3.	1	1	2	1
4.	1	1	1	1
5.	1	1	1	1
6.	1	1	2	1
7.	1	1	1	1
8.				
9.	2	3	1	2
10.	3	3		3
11.	2	2	1	2
12.	3	3	3	2
13.	1	1	1	1
14.	1	5	1	1
15.	1	2	1	1
16.	1	2	1	1
17.	1	1	1	1
18.	1	1	1	1
19.	1	1	1	1
20.	1	1	1	1
21.	2	2	1	1
22.	1	1	1	1
23.	1		4	1
24.	1	4	2	1

On the basic of the highest factor loadings on factors, few variables were selected to represent each factor. The variables selected are given below.

Variables to represent first factor

Variable 1. Size of holding
Variable 2. Gross cropped area irrigated
Variable 3. Gross area under high yielding variety of rice
Variable 4. Total costs of inputs (all crops)
Variable 5. Total rice input costs
Variable 6. Total agricultural income
Variable 7. Farm and irrigation equipment value

Variables to represent second factor

Variable 1. Yield per ha of irrigated rice
Variable 2. Total sale value of all crops

Variables to represent third factor

Variable 1. Yield per ha of irrigated rice/wheat
Variable 2. Yield per ha of unirrigated rice/wheat (in kg)
Variable 3. Yield per ha of unirrigated rice/wheat (value)

Variables to represent fourth factor

Variable 1. Farm and irrigation equipment value
Variable 2. Area irrigated of rice

Variables to represent fifth factor

Variables 1. Size of holding
2. Total costs of inputs
3. Value of farm and irrigation equipment

Variables to represent sixth factor

Variable 1. Area irrigated of the crop.

On the basis of values obtained for these selected variables to represent each of the six factors, all the 198 landholdings were ranked and grouped into clusters. Six groups of clusters were obtained and in terms of the nearness of association between landholdings, six clusters were obtained. A regional break up of clusters by the size of landholdings is prerented in Tables 6.16 to 6.21. Following features emerge from these tables.

Most of the landholdings get grouped under Cluster 5. 88 holdings are grouped in this cluster and it consists mainly of the medium and small holdings of rice and wheat. Only three large holdings of rice also get grouped here. The majority of small and medium holdings are rice holdings as is evident from the number of small rice holdings.

In this cluster three large rice holdings belong to Ganges Delta. This brings out that the large rice holdings in this region and small and medium rice holdings in the regions of West Coast, East Coast, North Eastern Peninsular Plateau are similar in their patterns of association.

Cluster 3 has a group of 53 landholdings. Most of the medium rice and wheat holdings are grouped here. There are seven large holdings under rice and only two large wheat holdings. All the large holdings under rice belong to the West Coast and East Coast regions. The medium holdings of wheat belong to Western U.P. Plain and Punjab, Haryana regions. This shows that the large landholdings of rice are similar in their pattern of association to the medium landholdings in wheat regions.

Cluster 4 has a grouping of 25 landholdings. Out of these, 23 are large holdings, most of them being rice landholdings. Only seven are large wheat holdings. Most large rice holdings belong to West Coast and East Coast regions. Two large holdings of North Eastern Peninsular Plateau are also grouped here. The large holdings of wheat belong mainly to Punjab Plains and Central irrigated area of Haryana. Only two large landholdings of Western U.P. Plains got grouped here. This cluster is different in the sense that it is a grouping of predominantly rice landholdings. But it is interesting to note that mostly landholdings in Punjab and Haryana are associated with the landholdings in West Coast and East Coast regions.

Cluster 6 has a grouping of only large holdings of rice and wheat. There are 12 holdings in all, out of which 8 belong to wheat and 4 belong to rice. Four of these large holdings belong to Punjab Plains, three to Malwa Plateau and one to Western U.P. Plains. Only four large landholdings under rice are

TABLE 6.16 **Regional Break up of Clusters by Sizes of Landholdings**

CLUSTER 1

Region	Rice Number of holdings				Wheat Number of holdings			
	Large	Medium	Small	Total	Large	Medium	Small	Total
1	2	3	4	5	6	7	8	9
West Coast	5	2	0	7				
East Coast	1	1	0	2				
Western U.P. Plains					2	0	0	2
Malwa Plateau					1	1	0	2
Punjab Plains and Central Irrigated area					2	2	2	6
Total	6*	3*	0*	9**	5*	3*	2*	10**

Total Number of Holdings—19

Note : * Refers to the number of holdings in each category.
** Refers to the number of holdings of all sizes.
Columns 2, 3, 4, 5 stand for rice holdings and 6, 7, 8, 9 for wheat.

TABLE 6.17 **Regional Break up of Clusters by Sizes of Landholdings**

CLUSTER 2

Region	Rice: Number of holdings Large	Rice: Number of holdings Medium	Rice: Number of holdings Small	Rice: Total	Wheat: Number of holdings Large	Wheat: Number of holdings Medium	Wheat: Number of holdings Small	Wheat: Total
1	2	3	4	5	6	7	8	9
East Coast	1	0	0	1				
Total	1*	0	0	1**				

Total Number of Holdings—1

Note ; * Refers to the number of holdings in each category.
** Refers to the number of holdings of all sizes.
Columns 2, 3, 4, 5 stand for rice holdings and 6, 7, 8, 9 for wheat.

TABLE 6.18 **Regional Break up of Clusters by Sizes of Landholdings**

CLUSTER 3

Region	Rice				Wheat			
	Number of holdings				Number of holdings			
	Large	Medium	Small	Total	Large	Medium	Small	Total
1	2	3	4	5	6	7	8	9
West Coast	0	7	5	12				
East Coast	4	10	3	17				
North Eastern Peninsular Plateau	3	1	0	4				
Western U.P. Plains					1	6	1	8
Malwa Plateau					0	4	0	1
Punjab Plains and Central Irrigated Area					1	10	0	11
Total	7*	18*	8*	33**	2*	17*	1*	20**

Total Number of Holdings—53

Note: * Refers to the number of holdings in each category.
** Refers to the number of holdings of all sizes.
Columns 2, 3, 4, 5 stand for rice holdings and 6, 7, 8, 9 for wheat.

TABLE 6.19 Regional Break up of Clusters by Sizes of Landholdings

CLUSTER 4

Region	Rice: Number of holdings				Wheat: Number of holdings			
	Large	Medium	Small	Total	Large	Medium	Small	Total
1	2	3	4	5	6	7	8	9
West Coast	6			6				
East Coast	9			9				
North Eastern Peninsular								
Plateau	2			2				
Western U.P. Plains					2	0	0	2
Central Irrigated Area of Haryana					4	0	0	4
Punjab					1	0	1	2
Total	17*	0*	0*	17**	7*	0*	1*	8**

Total Number of Holdings—25

Note : * Refers to the number of holdings in each category.
** Refers to the number of holdings of all sizes.
Columns 2, 3, 4, 5 stand for rice holdings and 6, 7, 8, 9 for wheat.

TABLE 6.20 **Regional Break up of Clusters by Sizes of Landholdings**

CLUSTER 5

Region	Rice				Wheat			
	Number of holdings				Number of holdings			
	Large	Medium	Small	Total	Large	Medium	Small	Total
1	2	3	4	5	6	7	8	9
West Coast	0	3	6	9				
East Coast	0	5	19	24				
North Eastern Peninsular Plateau	0	5	6	11				
Ganges Delta	3	6	9	18				
Western U.P. Plains					0	0	5	5
Malwa Plateau					0	4	8	12
Punjab Plains and Central Irrigated Area					0	2	7	9
Total	3*	19*	40*	62**	0*	6*	20*	26**

Total Number of Holdings—88

Note : * Refers to the number of holdings in each category.
** Refers to the number of holdings of all sizes.
Columns 2, 3, 4, 5 stand for rice holdings and 6, 7, 8, 9 for wheat.

TABLE 6.21 **Regional Break up of Clusters by Sizes of Landholdings**

CLUSTER 6

Region	Rice				Wheat			
	Number of holdings				Number of holdings			
	Large	Medium	Small	Total	Large	Medium	Small	Total
1	2	3	4	5	6	7	8	9
West Coast	1	0	0	1				
East Coast	2	0	0	2				
North Eastern Peninsular Plateau	1	0	0	1				
Western U.P. Plains					1	0	0	1
Malwa Plateau					3	0	0	3
Punjab Plains and Central Irrigated Area					4	0	0	4
Total	4*	0*	0*	4**	8*	0*	0*	8**

Total Number of Holdings—12

Note : * Refers to the number of holdings in each category.
** Refers to the number of holdings of all sizes.
Columns 2, 3, 4, 5 stand for rice holdings and 6, 7, 8, 9 for wheat.

grouped here, two belong to East Coast (Krishna Godavari delta in Godavari district, one to Chotanagpur Plateau in Bihar and one to West Coast in South Kanara district.

Cluster 2 has only one landholding belonging to East Coast in West Godavari district of Andhra. The size of holding of this large farmer is 33 hectares. Thus, it is isolated from all other landholdings.

An interesting feature that emerges from this pattern of clusters is that the large holdings of rice in most cases are similar to the medium and small holdings of wheat in this pattern of association. This indirectly brings out the fact that modern agricultural practices such as the adoption of high yielding variety have not penetrated deep enough even in the large and medium rice holdings. Another interesting feature is that most of the large and medium holdings in Ganges Delta Region are similar to the small and medium landholdings of wheat in Punjab, Haryana and Western U.P. Plains. It is not surprising to get this type of result, for even in the multiple regression results for the Ganges Delta Region, the results pertaining to inputs affecting yield per hectare were not significant at all. This partly reflects that the agricultural practices have remained almost the same over a long period and it is predominantly traditional agriculture that is practised in the region.[2] This holds in general for rice regions as compared to wheat regions. In the rice regions, in spite of new varieties of rice being introduced on a large scale particularly after 1970, the progress in terms of actual adoption of new techonlogy is still far behind than in wheat regions. There is a lot of difference between farm demonstrations carried out and the actual results obtained in the farmer's own holdings.

6.9 AN OVERVIEW—REGIONAL PATTERNS AND VARIATIONS IN AREA AND PRODUCTIVITY

When all the rice regions are considered together, the progress of adoption of high yielding variety is rather slow in most parts of India. In the seven villages studied in this analysis, except in parts of East Coast, HYV has not been adopted on a large scale, the main constraint being one of high costs associated with the adoption of HYV. One important factor which has not been covered by this study is the interest on capital borrowed. This was not included as the farmers did not give convincing figures. However, one can safely conclude that in most of the wheat regions in Punjab, Haryana, parts of U.P. interest on capital borrowed must be almost equal to the input cost they incur every year. Modern practices in wheat regions have been adopted on a large scale including in regions where it is cultivated on a small area. By contrast, in rice regions, the burden of credit did not appear to be heavy as per the information given by the farmers at the time of interviews. Thus, the high gross income from agriculture particularly in the better off wheat regions, should be taken with this qualification that it embodies in it, apart from other costs, a sizeable amount in the form of interest on capital borrowed. It appears that in wheat regions, farmers are willing to take risks associated with the adoption of HYV. This is true almost in all the categories of landholdings in Punjab and Haryana. A similar trend could not be seen in any of the rice villages covered in this study.

In the case of small and medium holdings, returns from agricultural production are highly correlated with area and yield of the crop. In the case of small holdings, the returns from agriculture are highly correlated with area under the crop. For medium holdings yields are correlated with returns from agriculture and for large holdings, size of holding and yield are correlated.

Finally, the results of cluster analysis bring out that the rice regions by and large are still behind the wheat regions in the adoption of modern practices of cultivation with some exception. It cannot be denied that in some rice areas as in parts of West and East Coast regions there has been a rapid shift over from traditional to high yielding varieties. But, there is a lot of difference between the demonstrations carried out on sample farms and the actual adoption of these varieties on farmer's own holdings in most parts of India. This is corroborated by the fact that the large and medium landholdings of rice regions got grouped with the low and medium landholdings of wheat regions.

REFERENCES

1. Refer Chapter 1 (*1.3*) for details.
2. Refer Chapter 5 for *details*.

7

Summary and Conclusions

This study has been an attempt in analysing the complex association among physical, economic and human factors that contribute to marked contrasts within and between different regions of the country in respect of agriculture. The study has been restricted to two major crops—rice and wheat. Such a study the author believes, is one of the first attempts to examine the spatial and temporal variations looking at the problem of spatial dimension at different area levels. Such an approach provides a realistic regional framework for evaluating potential of different crops in the country and to formulate the spatial strategies in conformity with the physical, economic and institutional characteristics of regions. Although the study is limited in nature, it shows the differential strategy required for agricultural development or for important crops that constitute the agricultural structure of the country. Obviously there are inherent limitations due to inadequate coverage in respect of data base and sample studies and yet it is noted that the study had added some new knowledge to reinforce the argument that the study of agriculture should be as close to ground reality as possible and generalisations would be meaningful at certain area levels depending on similarities in agro-climatic conditions, cropping patterns, productivity and so on.

The study covers spatial and temporal variations in area and yield responses of crops at three levels, namely, Macro, Meso and Micro. Macro level study covers main features of cropping pattern and crop yield at the national level for the year 1960-70. At the Meso level, state and intra-state variations have been analysed. Intra-State regions comprising groups of districts identified on the basis of similarity in natural factors and crop concentration, have formed the basis for evaluating the regional variations realistically. At the micro level, i.e. within homogeneous regions identified, the study has attempted to bring out the pattern of associations between landholdings of different sizes by regions which brings out inter-holding variations within and between regions. In the following section, a summary of the results emerging from the study made at Macro, Meso and Micro levels are presented.

7.1 CHANGES IN CROPPING PATTERN

At the national level, following results emerge. The percentage of area devoted to various crops has hardly changed between 1960-70. Crop area concentration has however changed in some states like Gujarat and Maharashtra where cotton and groundnut are gaining importance. Tamil Nadu, Karnataka and Andhra are three other states where groundnut is an important crop. These five states together accounted for 80 per cent of national area under groundnut. However food crops predominate in all the states except Gujarat where area under groundnut and cotton is higher than under bajra which is next in importance to these crops in terms of percentage of cropped area. In 1970-72 food crops accounted for 75 per cent of total cropped area at the national level of which, cereals accounted for 61.2 per cent and pulses 13.8 per cent. Rice and wheat occupy the first two places. In terms of percentage growth rate in area under different crops between 1960-63 and 1967-70, there are marked inter-state variations In some states like Orissa, area under all crops increased in an attempt to bring more area under tillage. In almost all the states the area under food crops increased during this period. However, it is interesting to observe that Rice area has increased in Punjab-Haryana while wheat area has increased in Assam, Orissa, West Bengal and Gujarat between 1960-63 and 1967-70. Wheat cultivation is now penetrating into the hitherto rice growing regions.

Finally, it was observed that during 1960-63 and 1967-70, in most states increase in output is significant while yield growth rates are not very significantly different from what they were earlier, especially for food crops like rice and wheat in the traditional rice and wheat cultivating regions. However, exceptions to this are the core areas of wheat in Punjab-Haryana where the yields have risen. Interestingly, yield variations of crops are also more marked in the areas where rice or wheat are of more recent origin. For example in West Bengal, wheat yield increased significantly. In Punjab, rice yields improved significantly, in the period 1967-70. In Punjab, yield of all crops increased and in this respect this region is exceptional. The overall trend of yield in this period is however, one of decline.

7.2 CORE REGIONS AND THEIR SHARE TO THE NATIONAL TOTAL

Regional dimension can be better understood and evaluated in terms of smaller administrative units than the states which are of varied size, and often too heterogeneous within the data constraints and possibilies of providing regional dimension to agriculture at different levels of areas. For this purpose, the district was taken as an ideal unit for data computation and analysis and regional grouping based on the concept of homogeneity in one or more factors related to agriculture. The spatial and temporal variations in the pattern of distribution of rice and wheat in India have been analysed using districtwise data for four time blocks, namely 1960-63, 1963-66, 1966-68 and 1970-72. The core and periphery areas of rice and wheat are marked off on a map as the initial step in demarcating regions most favourable to these crops in India. The importance of rice and wheat in each region was brought out by working out the proportion of area under rice and wheat to gross cropped area of the districts and in turn for different regions. The districts are ranked and this brought out the relative importance of districts. In the case of rice regions, during 1960-63, the cropping pattern showed that the percentage of area under rice to total cropped area varied from 30 to 70 per cent and in 1970-72 also it varied between 30 to 70 per cent in the case of West Coast, East Coast, North Eastern Peninsular Plateau and Lower Ganges Plain. The corresponding national average was 22 per cent in 1960-63 and 25 per cent in 1970-72. During

1963-66 and 1966-68, the proportion of area under rice to total cropped area remained almost the same. However, during 1970-72 area under rice to total cropped area increased by infilling in the same area with exception of Punjab-Haryana area where there has been substantial increase in rice area. Though rice dominates the cropping pattern in areas of over 100 inches of rainfall and above, in some regions of 70-100 inches of rainfall also rice occupies a significant place in the crop pattern.[1] In the case of wheat one finds that during 1960-63, higher area concentration (above national average of 7.2 per cent of total cropped area) prevailed in the plains of Western Uttar Pradesh, Punjab Plains, Central Irrigated area in Punjab and Haryana, parts of Peninsular Plateau and North Western transitional region. There was an increase of 7.8 per cent in the percentage of area under wheat to total cropped area at the all India level during 1970-72 Seventy six wheat districts were placed above the national average in 1970-72, whereas in 1960-63, one hundred six districts ranked above the national average. In this respect, wheat and rice differ in that in the case of wheat, there has been a larger increase in the area under wheat in relation to total cropped area whereas for rice it increased by only 3 per cent which is much lower than 7.8 per cent achieved in the case of wheat in 1970-72. In contrast to rice which is widely distributed extending from the Coastal Plains to parts of Ganges Plain, wheat is dry and highly concentrated in few parts. In the study of spatial variations in pattern of yield growth rates in rice and wheat, it is observed that in the case of rice 126 districts ranked above the national average yield growth rate in 1970-72 with the average of 1963-66 taken as base,[2] whereas in 1970-72 with the average of 1960-63 taken as base, only 109 districts ranked above the national yield growth rate. Thus the impact of modern methods of cultivation seem to be unevenly spread in the case of rice. Whatever the increase has been, it is only since 1966. In contrast, in 1963-66 with average of 1960-63 taken as base, national average growth rate in rice was itself negative (—0.2%) and 171 districts had negative yield growth rate. This was a period characterised by low rice yields all over. Thereafter, rice yields have increased significantly in areas where it is grown a new, for example in Punjab-Haryana, as compared to the increase, in the traditional rice regions. The spatial patterns and variations in wheat yield reveal that growth rate in wheat has been extensive covering most parts of traditional wheat growing areas as well as in areas brought under wheat cultivation anew. The national average growth rate in wheat yield was 62.9 per cent in 1970-72 with reference to the average of 1963-66 as base and seventy districts ranked above the national average. This shows a contrast with rice where national average yield growth rate was only 11.7 per cent in this period. However, in the case of growth rates in wheat yield it was observed that both with 1960-63 as base and 1963-66 as base, the variations in the spatial patterns of yield growth rates were significant. Thus yield increases were a continuing feature in most areas of wheat growing regions.[3] As in the case of rice, in wheat also, during 1963-66 with 1960-63 as base, national average was negative for wheat yield growth rate. Seventy two districts had negative yield growth rate.

In the analysis of overall pattern of rice and wheat in the agricultural landscape, it was observed that the core regions[4] of rice contributed 71.10 per cent of area to area under rice in the country as a whole in 1960-63, and in 1970-72, they contribute 71.12 per cent which is hardly different. The same trend is observed even in the case of their contribution to rice production in the country as a whole. The core regions[5] share to national production was 73.20 per cent during 1960-63 and 73.14 per cent in 1970-72. Thus, their contribution has remained almost the same. In area concentration, North Eastern Peninsular Plateau had highest share in 1960-63 and 1970-72. In production the highest share was accounted for by East Coast region. The average yield was the highest in West Coast region followed by East Coast region. In 1960-63, the national average yield of rice was 991 kgs per hectare and in 1970-72, it was 1132 kgs per hectare. Upper Ganges Plain had the lowest average yield.

In the case of core regions of wheat, their contribution to area under wheat in the country as a whole was 70.06 per cent in 1960-63 and 87.75 per cent in 1970-72. This is a significant increase. This is in contrast to the results that we observed in the case of core regions of rice. The percentage share of core regions of wheat in national production was 73.46 per cent in

1960-63 and 77.08 per cent in 1970-72. In the case of core rice regions their contribution to production had remained the same between 1960-63 and 1970-72. There was a significant increase in national average yield of wheat between 1960-63 and 1970-72. It rose from 845 kgs per hectare in 1960-63 to 1343 kgs per hectare in 1970-72. Only Punjab Plains (including Central Irrigated area of Haryana) had above national average yield level. In 1960-63 this region had an average yield level of 1298 kgs per hectare and in 1970-72, it had an average yield of 1898 kgs per hectare which is a significant increase.[6] A general obvious conclusion drawn from this spatial analysis was that rice and wheat regions are still mutually exclusive except in parts of Madhya Pradesh, Jammu & Kashmir and Lower Ganges Plains both in terms of cropping pattern and relative share to total area and production in the country as a whole. The core regions provide a realistic framework for concentrated effort in agricultural development particularly of rice and wheat.

7.3 MEASUREMENT OF ASSOCIATION AMONG FACTORS INFLUENCING THE CULTIVATION OF RICE AND WHEAT

These findings and the regional framework provided the basis for the measurement of association among factors contributing to spatial concentration and spread of rice and wheat and variations in their responses to high yielding variety in the core regions. Multiple linear regression analysis was attempted in order to analyse the influence of certain factors on area and yield of Rice/Wheat in the core region. In one model area was taken as the dependent variable, with five independent variables, namely, growth rate in gross cropped area, variation in gross area irrigated as a percentage of gross cropped area, variation in percentage of area irrigated of the crop to total area under the crop, growth rate in yield in rupees per hectare and finally growth rate in yield in kilograms per hectare. In the other model yield in kilograms/hectare was taken as the dependent variable and independent variables were, fertiliser consumption per hectare, tractors per thousand hectares, pump sets and engines per thousand hectares, percentage of rice/wheat area irrigated, percentage of area under high yielding variety of rice/wheat to total area under rice/wheat and finally price of the crop (Rupees per Kg.).

7.4 SUMMARY OF SALIENT FEATURES OF ASSOCIATION

Following is a summary of the main results.[7]

The variables behaved more uniformly in wheat regions than in rice regions. In 1970-71 in all wheat regions except Punjab Plains proportion of wheat area irrigated has an important influence on wheat yields. In the case of rice regions, this factor has emerged as significant only in North Eastern Peninsular Plateau. This is obvious, for, in this region, rice cultivation is partly rainfed and partly irrigated. Another observation made was that fertilizer consumption per hectare and price of the crop have influenced rice yields The influence of these two variables was negative. Also the influence of pump sets, tractors on rice yields is not significant. This factor emerges significant in its influence on wheat yields.[8]

In the North Western Transitional region (Rajasthan), all the variables together explain 76 per cent of total variation in wheat yield. This is followed by Punjab plains (62 per cent), North Eastern Peninsular Plateau (50 per cent) and North Peninsular Plateau (35 per cent) respectively. In the wheat regions together, the variables explained 69 per cent of total variation in wheat area in 1970-71 and in the case of all rice regions combined, they explained 40 per cent of total variation in rice yields. However in the regions of West Coast and East Coast, the explained variations by all the variables taken together is 92 per cent and 79 per cent respectively. Except in Lower and Upper Ganges Plain region, most of the results are significant in all other rice regions.

However when on looks at the relevance of all the variables in explaining crop yields, the variables in this model have explained wheat yield variations more significantly than rice yield variations. All said and done, rice cultivation in most parts of India is still highly correlated with the weather factor, namely the timely arrival of rains. But for inter-regional analysis of this type, the element of crop-weather relationship is taken for granted and the element is embodied in it.

In order to probe deeper into the variations in rice and wheat cultivation and the pattern of association within and between regions of rice and wheat, field enquiry was conducted in sample holdings spread over different regions of rice and wheat. Purposive sampling became necessary and there were selected seven rice villages and four wheat villages. The rice villages were selected from the regions of West Coast, East Coast, North Eastern Peninsular Plateau and Ganges Delta. The wheat villages were selected from Upper Ganges Plain regions, Malwa Plateau and Punjab Plains region (including Central Irrigated area of Haryana). In all, 198 landholdings were covered in the study. 72 landholdings were covered in wheat regions and 126 in rice regions. The results were based on primary data collected through sample survey with the help of a questionnaire. From each village eighteen households were interviewed, six each from large, medium and small holdings. Data pertaining to 24 variables were generated and these were converted into percentages for inter-regional and intra-regional comparisons. The observations based on these percentages are as follows. When all the rice regions are considered together, the process of adoption of high yielding variety is rather slow in most parts of India. In the seven villages studied in this analysis except in parts of East Coast, HYV has not been adopted on a large scale, the main constraint being one of costs associated with the HYV. The most important findings of this survey based on sample are the patterns of similarities and dissimilarities in cluster formation of holdings of large, medium and small categories across the rice and wheat regions and variations between different sizes of landholdings within a region. The pattern of association is such that within a region, the large holdings got scattered over space in such a way that large and medium holdings belonging to one region showed their affinity to the large holdings in another region. Such permutations and combinations do really occur and one finds interesting patterns of association emerging. In this analysis, an attempt was made to bring out such affinities which clearly highlight the variations in the characteristics of holdings within the region. This was done with the help of cluster analysis.

7.5 SALIENT FEATURES OF THE PATTERN OF CLUSTERING

Following were the main results of cluster analysis. The pattern of clusters obtained was such that the larger holdings of rice in most cases are similar to the medium and small holdings of wheat. This indirectly brings out the fact that modern agriculture practices such as adoption of high yielding variety has not penetrated deep enough even in the large rice holdings. Another interesting feature of cluster analysis was that all the large landholdings in West Bengal get grouped with the small and medium wheat holdings of Upper Ganges region and Punjab Plains. In fact an interesting feature is that there is not much difference between large, medium and small holdings in this region for, all the holdings got grouped in Cluster 5. Three holdings are in large category, six are in the medium category and nine in small category.[9] In fact only the three large holdings of West Bengal came in that category. In this respect, West Bengal stands out in contract from other rice regions, because, in this cluster only the medium and small holdings of rice and wheat were grouped. The large holdings of West Bengal thus got grouped with the medium and small holdings of rice and medium and small holdings of wheat.

Another interesting pattern emerges in Cluster 6.[10] Here, only the large holdings get grouped. The large holdings of rice belonged mainly to East Coast region and only one large holding belonged to West Coast. The large holdings of wheat belong mostly to Punjab and two to Malwa Plateau. This gives a pattern which is entirely different from the pattern emerging in other clusters. The large holdings of rice in these regions are similar in their characteristics to the large holdings of wheat regions. Cluster 2 has only one large holding belonging to West Coast region. This particular sample had an average size of holding of nearly 30 hectares which is distinct in its characteristics as compared to all the rice regions and all wheat regions. So this is the only holding emerging in this cluster.

In general, for the 198 holdings, for the small and medium holdings returns from agricultural production are highly correlated with area and yield of rice/wheat. In the case of small holdings, returns from agriculture are highly correlated with area under rice/wheat. For medium holdings yields are correlated with area under the crop. For large holdings, size of holdings and crop yields are highly correlated.

An important factor which is not covered in this study is the Interest on Capital borrowed in calculating costs associated with agricultural operations. This was not considered, as reliable information could not be obtained from the farmers. However, on the basis of information collected, it could be concluded that in wheat regions of Uttar Pradesh and Punjab-Haryana, the investment in agriculture is very heavy and in their case, the interest on capital borrowed was almost equal to their input costs incurred every year. The burden of credit did not appear to be heavy in the case of rice holdings, as per the information collected. Therefore one should not forget that while wheat regions are better off in terms of agricultural income, they are worse off if one took into account the enormous cost involved in modernising agriculture. On the whole, it appears that farmers in the wheat regions are willing to take risks associated with the transformation of agriculture, as compared to farmers in rice regions as revealed by this study. A general conclusion that can be drawn from this study is that rice regions by and large are still behind the wheat regions in the adoption of modern agricultural practices with some exception where there has been a shift over from traditional varieties to high yielding varieties. An important question that arises from this study is, why have rice yields not increased in areas of high concentration whereas wheat yields have increased all over? Perhaps part of the reason lies in lack of initiative of the farmer to introduce efficient methods of cultivation in some of the rice regions like Orissa and West Bengal. In Bengal and Uttar Pradesh, ground water supplies should be harnessed to solve the irrigation problems. There is therefore, a wide gap between the revolution achieved in sample farm-holdings and performances in the farmer's own holdings. Apart from factors like costs associated with modernization, the reluctance of the farmer arising from the inequalities relating to the supply of inputs, which arrive either too late or, are in insufficient quantity at high prices, was also responsible for this gap. Adequate involvement of the farmer is a prerequisite for any revolution in agriculture in India.

The number of samples is relatively small taking into consideration the large size of the country. Yet, it is hoped that the results provide new insight and bring out the challenging areas of research still unexplored at the regional levels. To that end, this is an attempt to bring in inter-disciplinary approach, combining physical base with human response.

REFERENCES

1. Refer Figures 1 to 4 in Chapter 4 for details.
2. Refer Figures 9 and 10 in Chapter 4 for details.
3. Refer Figures 12 and 13 in Chapter 4 for details.
4. Refer Appendix 'B' at the end of Chapter 5 for core region.
5. See Table 4.1 in Chapter 4 for details.
6. Refer Table 4.2 in Chapter 4.
7. Refer Chapter 5 for details.
8. It must however be mentioned that in the combined effect of modernisation, it is difficult to isolate the impact of different inputs separately.
9. See Table 6.16 in Chapter 6.
10. Refer Table 6.17 in Chapter 6.

Bibliography

1. Articles

Abraham, T.P. and S.K. Rakheja (1967) "An Analysis of Growth of Production of Rice and Wheat Crops in India", *Indian Journal of Agricultural Economics*, 22 (3), July-September, pp. 1-15.

Agarwal, G.D. (1958) "Output-Input Relations in Indian Agriculture", *Indian Journal of Agricultural Economics*, 13 (3), July-Sept, pp. 26.3.

Bashir Desai and Thingalaya, N.K. (1965) "Irrigation factor and Yield variability in rice growing districts of India", *Indian Journal af Agricultural Economics*, July-September.

Bhardwaj, K.R. and R. Bhardwaj (1965) "An Activity Analysis Approach to Measure Productivity Efficiency in Agriculture (A Case Study of Ahmednagar and Nasik Districts)", *Indian Economic Journal*, 12 (3), January-March, pp. 211-231.

Bhattacharjee, J.P. (1949) "Mechanisation of Agriculture in India—Its Economics", *Indian Journal of Agricultural Economics*, 4 (1), March, pp. 121-22.

Chatterji A , (1966) "A Study of Agricultural Growth during 1950-63 in India", *Indian Journal of Agricultural Economics*, 21 (4), October-December, pp. 192-200.

Chauhan, K.K.S. and N.L. Agarwal (1969) "Changing Pattern and Magnitude of Capital Investment in Agriculture in Sri Ganganagar District of Rajasthan", *Indian Journal of Agricultural Economics*, 24 (4), October-December, pp. 136-137.

Dantwala, N.L. and W.B. Donde (1949) "The Uneconomic Cultivator", *Indian Journal of Agricultural Economics*, 4 (2), August, pp. 9-48.

Dantwala, N.L. (1959) "Small Farmers, Not Small Farms",

Indian Journal of Agricultural Economics, 14 (3), July-September, pp. 58-59.

Desai, D.K. and D.A. Patel (1968) "Developing Hybrid Seed Enterprise—An Induced Growth of High-Yielding Varieties Programme in India", *Indian Journal of Agricultural Economics*, 23 (4), October-December, pp. 113-124.

Desai, Gunwant N. (1971) "Some Observations and Economics of Cultivating High-Yielding Varieties of Rice in India", Paper presented at the Rice Policy Conference organised by the International Rice Research Institute, Los Banos, Philippines, May 9 to 13.

Ghosh, N.G. and N. Bandyapadhyay (1964) "A Comparitive Study of Some Aspects of Agricultural Development in two States of Eastern Zone—West Bengal and Orissa", *Indian Journal of Agricultural Economics*, 19 (1), January-March, pp. 212-219.

Giri, R., A.V.K. Shastri and D.S. Somayajulu (1966) "Components of Crop Output Growth in India", *Indian Journal of Agricultural Economics*, 21 (4), October-December, pp. 183-191.

Gohain, D. (1968) "Economic Aspects of High Yielding Varieties Programme", *Indian Journal of Agricultural Economics*, 23 (4), October-December, pp. 73-79.

Gopalakrishnan, H.D. and T. Ramakrishnan (1964) "The Regional Variations in Agricultural Productivity in Andhra Pradesh", *Indian Journal of Agricultural Economics*, 19 (1) January-March, pp. 227-236.

Jai Krishna (1967) "Evaluation of Intensive Agricultural District Programme : A Case Study for Wheat in Ludhiana District", *Indian Journal of Agricultural Economics*, 22 (3), July-September, pp. 82-88.

John, P.V. (1965) "Representativeness of Relative Area Output of Sugarcane and Rice to changes in their Relative Prices in Uttar Pradesh 1954-63", *Indian Journal of Agricultural Economics*, 20 (1), January-March 1965, pp. 40-47.

Kahlon, A.S. and A.C. Sharma (1967) "Strategy of Mechanisation and Improvement in Indian Agriculture", *AICC Economic Review*, 18 (15), pp. 31-34.

Kaul, J.L. (1966) "Value Productivity Growth of Some Important Crops in the Punjab", *Indian Journal of Agricultural Economics*, 21 (4), October-December, pp. 200-204.

Maitrá, T. and Rina Roy (1964) "Regional Variation in Yield per Acre of Major Crops in India", *Indian Journal of Agricultural Economics*, 19 (1), January-March, pp. 168-176.

Majid, Abdul (1963) "Crop Pattern and Size of Cultivated Holdings", *Indian Journal of Agricultural Economics*, 18 (1), January-March, pp. 97-101.

Minhas, B.S. and A. Vaidyanathan (1965) "Growth of Crop Output in India 1951-54 to 1958-61—An Analysis of Component Elements". *Journal of the Indian Society of Agricultural Statistics*, 17 (2), pp. 230-252.

Minhas, B.S. and T.N. Srinivasan (1966) "New Agricultural Strategy Analyses", *Yogana*, 10 (1), January 26, pp. 20-24.

Mukherjee, P.K. (1970) "The HYV Programme—Variables that Matter", *Economic and Political Weekly*, 5 (13), Review of Agriculture, March 28, 1970, pp. A-15, A-22.

Mathiah, C. (1971) "The Green Revolution—Participation by Small Vs. Large Farmers", *Indian Journal of Agricultural Economics*, 26 (1), January-March, pp. 53-66.

Misra, V.N. (1971) "Growth of Crop Output in Gujarat : A Component Analysis", *Anvacak* (1), June, pp. 1-15.

Ojha, Gyaneshwar (1969) "Changes in the Pattern of Capital Formation in Agriculture", *Indian Journal of Agricultural Economics*, 24 (4), October-December, pp. 138-139.

Panse, V.G. (1952) "Trends in Area and Yields of Principal Crops in India", *Agricultural Situation in India*, 7 (3), June, pp. 144-148

Panse, V.G. (1959) "Recent Trends in the Yield of Rice and Wheat in India", *Indian Journal of Agricultural Economics*, 14 (1), January-March, pp. 11-58.

Parikh, Ashok (1966) "Statewise Growth-Rate in Agricultural Output—An Econometric Analysis", Artha Vijnana, 8 (1), March, pp. 1-52.

Parikh, Ashok (1964) "Cropwise Districtwise Production Functions", *Indian Journal of Agricultural Economics*, 25 (1), January-March, pp. 43-58.

Patel, S.J. (1953) "Trends of Unit Yields of Foodgrains in India, 1936-50", *Indian Journal of Agricultural Economics*, 34 (132), Part I, July, pp. 49-59.

Parikh, Ashok (1969) "Complementarity between Irrigation and Fertilisers in Indian Agriculture", *Indian Journal of Agricultural Economics*, 24 (3), July-September, pp. 1-18.

Prasad, A.G. (1969) "Capital Investment in Agriculture : A Study in Regional Variations", *Indian Journal of Agricultural Economics*.

Radhakrishna, D. (1964) "A Study of Regional Productivities of Agricultural Inputs", *Indian Journal of Agricultural Economics*, 19 (1), January-March, pp. 237-242.

Ramalingam, C. (1963) "Some Economic Aspects of Cropping Pattern", *Indian Journal of Agricultural Economics*.

Rao, C.H. Hanumanatha (1963) "Growth of Agriculture in the Punjab during the Decade 1952-62", *Indian Journal of Agricultural Economics*, 20 (3), July-September, pp. 20-32.

Rao, C.H. Hanumanatha (1964) "Alternative Explanations of the Inverse Relationship between Farm Size and Output per Acre in India", *Indian Economic Review*.

Rao, C.H. Hanumanatha (1970) "Farm Size and Credit Policy", *Economic and Political Weekly*, 5 (52), Review of Agriculture, December 26, pp. A 157-A 162.

Rao, C.H. Hanumanatha (1965) "Growth of Agriculture in Punjab during the Decade 1952-62", *Indian Journal of Agricultural Economics*, July-September

Saxena, B.C. (1962) "Problems of Marketable Surplus in India", Agricultural Situation in India, 16 (12), March, pp. 1212-1214.

Shah, C.H. (1956) "Crop Variations 1943-1954 : A study in crop pattern change for farmers in different size groups of holding", *Indian Journal of Agricultural Economics*, 11 (3), July-September, pp. 41-58.

Shah, S.H. (1963) "Cropping Pattern in Relation to Irrigation", *Indian Journal of Agricultural Economics*, 18 (1), January-March, pp. 154-160.

Singh, Roshan and B.B. Singh (1972) "Mechanisation of Farming in India" (Paper submitted at the Seminar on Farm Mechanisation held on July 24-26, 1970 at Agro-Economic Research Centre, Vallabh Vidyanagar), *Seminar Series-IX*, *Indian Society of Agricultural Economics*, Bombay, 1972.

Subbaraju, V. (1949) "Mechanisation of Indian Agriculture",

Indian Journal of Agricultural Economics, 4 (1), March, pp. 104-120.

Tambad, S.B. (1965) "Spatial and Temporal Variations in Agricultural Productivity in Mysore State", *Indian Journal of Agricultural Economics*, 20 (4), October-December, 1965, pp. 37-45.

Thamarajakshi, R. (1967) "Agricultural Growth in Madras State 1949-50—1962-63", Agricultural Situation in India, 22 (9), December, pp. 1003-1006.

Venkateswarlu, V. (1965) "Growth in Agricultural Output in Andhra Pradesh during the period 1952-53 to 1961-62", *Arthaniti*, 8 (2), July, pp. 157-170.

Venkataram, J.V. and R. Ramanna (1968) "Economic Aspects of High Yielding Varieties of Crops Hybrid Jowar", *Indian Journal of Agricultural Economics*, 23 (4), October-December, pp. 134-138.

Garg, J.S. (1964) "Variation Studies in the Agricultural Development and Productivity in the Eastern and Western Regions of Uttar Pradesh", *Indian Journal of Agricultural Economics*, 19 (1), January-March, pp. 193-197.

Gupta, Amiya K. (1963) "Inter-State Differences in Cropping Pattern and Productivity", *Indian Journal of Agricultural Economics*, 18 (1), January-March, pp. 24-31.

2. Books and Reports

Ansari, Resin (1968) *Economic of Irrigation Rates—A Study in Punjab and Uttar Pradesh*, Agricultural Economics Research Centre, University of Delhi, Asia Publishing House, Bombay.

Arora, R.C. (1976) *Development of Agriculture and Allied Sectors—An integrated Area Approach.*

Bhat, L.S and Das, B.N. (1967) *Regional Variations in National Environment and the Identification of Regional Physical Complexes for the Evaluation of Agricultural Land in India.*

Buck, J.L. (1956) *Land Utilisation in China.*

Economic and Statistical Organisation, Haryana (1969-70) *Changing Structure in Agriculture in Haryana*—A case study of the impact of green revolution.

Government of India, Planning Commission (1957) Studies in

the Economic of Farm Management 1954-55 in (i) Bombay (ii) Madras (iii) Madhya Pradesh (iv) Punjab (v) Uttar Pradesh and (vi) West Bengal, Directorate of Economic and Statistics, Ministry of Food, Agriculture, Community Development and Cooperation, New Delhi.

Government of India, Planning Commission (1969) Report on the High Yielding Varieties Programme (Kharif 1967-68), Directorate of Economics and Statistics, Ministry of Food, Agriculture, Community Development and Cooperation, Government of India, New Delhi.

Gurdev Singh and Ojha, B.S. (1967) *Agricultural Land Use in Punjab (A Spatial Analysis).*

Indian Council of Agricultural Research (1971-72) *All India Coordinated Agronomic Experiments Scheme.*

Indian Council of Social Sciences Research (1975) *A Survey of Research in Economics*, Vol. III, Agriculture (1).

National Council of Applied Economic Research (1964) *Factors Affecting Fertilizer Consumption (Problem & Policies).*

National Council of Applied Economic Research and the Fertilizer Association of India (1974) (A joint Publication) *Fertilizer Use on Selected Crops in India.*

Parikh, S. and Srinivasan T. (1974) *Optimum Requirement of Fertilizers for the Fifth Plan Period.*

Parthasarathy, Gogula (1971) *Agricultural Development and Small Farmers.*

Programme Evaluation Organisation, Govt. of India (1967) *Study on the Use of Fertilizers & Agricultural Production*, Planning Commission.

Rao, C.H. Hanumanatha (1965) *Agricultural Production Functions, Costs & Returns in India*, Asia Publishing House, Bombay.

Spate, O.H.K. (1965) *India and Pakistan.*

United News of India record (1975) *Aspects of Agriculture in India*, Editored by G. Mirchandani.

Vyas, V.S.D.S. (et al.) (1969) *Significance of the New Strategy of Agricultural Development for Small Farmers—A Cross Section Study of Two Areas of Gujarat*, Agro Economic Research Centre (for Gujarat and Rajasthan), Sardar Patel University, Vallabh Vidyanagar.

3. Sources of Published Data

Government of India, Ministry of Agriculture, Directorate of Economics and Statistics, *Growth Rates in Agriculture—* 1949-50 to 1964-65.

———*Indian Agricultural Statistics*, 1961, 1962-63 (Detailed Tables, Vol. II.

———*Indian Agricultural Statistics* (1963-64 and 1964-65), Detailed Tables.

———*Season and Crop Report Estimates*, All Available Volumes.

———*Statistical Abstracts of States*, Agricultural Situation in India, Nov. 1970, June 1970, Nov. 1971, March 1971, Feb. 1972, March 1972, July 1973, Sep. 1973.

———*Farm Harvest Prices of Principle Crops*, 1954—55 to 1964-65 & 1965-66 to 1970-71.

———*Bulletin of Rice Statistics in India*, 1974.

———*Bulletion of Wheat Statistics in India.*

Fertilizer Association of India, 1970-71, & 1972-73—Fertilizer Statistics.

Index